BrightRED Study Guide

CfE HIGHER

PHYSICS

John Taylor

BrightRED
PUBLISHING

First published in 2015 by:
Bright Red Publishing Ltd
1 Torphichen Street
Edinburgh
EH3 8HX

Reprinted with corrections in 2016

A CIP record for this book is available from the British Library.

ISBN 978-1-906736-67-5

With thanks to:
PDQ Digital Media Solutions Ltd, Bungay (layout), Project One Publishing Solutions Ltd (copy-edit).

Cover design and series book design by Caleb Rutherford – e i d e t i c.

Acknowledgements
Every effort has been made to seek all copyright-holders. If any have been overlooked, then Bright Red Publishing will be delighted to make the necessary arrangements.

Permission has been sought from all relevant copyright holders and Bright Red Publishing are grateful for the use of the following:

VectorState/Reshetnyova Oxana (p 6); VectorState (p 6); Microstock Man/Shutterstock.com (p 14); Cult content (CC BY-ND 2.0)[1] (p 17); Caleb Rutherford e i d e t i c (p 21); Neil Williamson (CC BY-SA 2.0)[2] (p 22); massawfoto/iStock.com (p 22); Image licensed by Ingram Image (p 26); mirabo11/freeimages. com (p 26); Image licensed by Ingram Image (p 26); NASA (p 26); rypson/iStock.com (p 27); manahan/ freeimages.com (p 27); Ingy the Wingy (CC BY-SA 2.0)[2] (p 32); Apollo 11, NASA (p 32); Caleb Rutherford e i d e t i c (p 32); Image licensed by Ingram Image (p 37); VectorState/Laschon Maximilian (p 39); NASA (p 43); Title page of 'Principia', first edition (1687) (public domain) (p 44); Image Editor (CC BY 2.0)[3] modified from NASA's Photojournal Home Page graphic released in October 2007, catalog ID: PIA10231 (NASA/JPL) (p 45); NASA (p 46); Two images from NASA Goddard Space Flight Center (CC BY 2.0)[3] (p 47); Xorx (CC BY-SA 2.5)[4] (p 50); NASA Goddard Space Flight Center (CC BY 2.0)[3] (p 51); NASA, ESA, the Hubble Heritage (STScI/AURA)-ESA/Hubble Collaboration, and A. Evans (p 52); NASA/ JPL/Caltech/Harvard-Smithsonian Center for Astrophys (p 52); NASA, retrieved from universe-review. ca (p 53); NASA (p 54); NASA Goddard Space Centre (p 55); restyler/iStock.com (p 58); Image licensed by Ingram Image (p 59); monticelllo/iStock.com (p 71); scanrail/iStock.com (p 71); Department for Business, Innovation and Skills (CC BY-ND 2.0)[1] (p 81); Maximilien Brice, CERN (p 81); INeverCry (public domain) (p 81); Pcharito (CC BY-SA 3.0)[5] (p 81); Henry Mühlpfordt (CC BY-SA 2.0)[2] (p 91); Emoscopes, Sakurambo (CC BY-SA 3.0)[5] (p 97); Abteilung Öffentlichkeitsarbeit (CC BY-SA 3.0)[5] (p 97); Alamy (p 103); setixela/iStock.com (p 111); bzuko22 (p 119); bzuko22 (p 119); NASA Goddard Space Flight Center (CC BY 2.0)[3] (p 119).

(CC BY-ND 2.0)[1]   http://creativecommons.org/licenses/by-nd/2.0/
(CC BY-SA 2.0)[2]   http://creativecommons.org/licenses/by-sa/2.0/
(CC BY 2.0)[3]   http://creativecommons.org/licenses/by/2.0/
(CC BY-SA 2.5)[4]   http://creativecommons.org/licenses/by-sa/2.5/
(CC BY-SA 3.0)[5]   http://creativecommons.org/licenses/by-sa/3.0/

Printed and bound in the UK by Charlesworth Press.

CONTENTS

INTRODUCTION

INTRODUCING CFE HIGHER PHYSICS

WHY PHYSICS?

- Studying **physics** deepens our understanding of our **world** and the **universe** in which it exists.
- Understanding physics is fundamental to a deeper understanding of **all science**.
- Physics studies the **requirements** we need to improve everyday **life** through to the **advances** in the exploration of **space**, and we can consider the many **applications** that have been developed as a result of the discoveries of the **laws of physics**.
- Physics has been developed as a result of **practical experimentation** and **theoretical thinking** and you will also develop these **skills** as you do this course.
- The study of physics ranges in **scale** from the smallest discoveries in **particle physics** through to the incredible magnitude of the **universe** itself.
- Modern **technology** exists and develops as a result of our understanding of physics.

WHY DO I NEED THIS BOOK?

This book provides the content of the CfE Higher Physics syllabus in a **concise** and **attractive** format. Each double page spread presents a new topic which will make learning accessible and digestible. **Key ideas** are illustrated in **colour**. **Key terms** are highlighted to emphasise information that you need to understand to be able to handle the tricky 'explain' type of examination question. The **graphical** presentations allow **experimental concepts** to be understood through **visual learning**.

By studying this book you should be re-inforcing and extending your **knowledge and understanding** of the concepts of physics. By studying worked examples, practising problems and following the '**Things to do and think about**' features, you will improve your theoretical and **problem-solving** skills. You will be expected to develop **investigative**, **research** and **experimental** skills as you progress through the Higher course and this will deepen your understanding of physics.

As you progress through the course, you will find that the **theory** and **concepts** are based on, and illustrated by, experimental activities. Physics involves interaction between theory and practice and pure physics concepts are illustrated with applications.

SYLLABUS

This book follows the **mandatory units** of the syllabus:
- **Researching Physics (Higher)**   Half unit
- **Physics: Our Dynamic Universe (Higher)**   Full unit
- **Physics: Particles and Waves (Higher)**   Full unit
- **Physics: Electricity (Higher)**   Half unit

Skills

Our chapter entitled **Skills** includes extremely useful advice for dealing with the **Researching Physics** unit. This unit can be approached in various ways and you may have choice in your areas of research. The information given will help guide you in the **skills** that you should seek to develop, how to carry out your **research**, and in how to **present findings** to a Higher standard.

Along with **knowledge and understanding** of the main units, you will use your **practical experience**, **investigations** and **problem-solving** situations to **develop skills** of scientific inquiry, investigation and analytical thinking. Your developed skills should help you to consider the applications of these units on our lives, as well as the implications on society and the environment.

ONLINE

Online and video **links** are also provided throughout the book. These are intended to **illustrate a point**, further **understanding** or provide **additional interest** to the topic.

DON'T FORGET

'**Don't forget**' reminders will draw your attention to points being made, deepen your understanding, and provide reminders which will be helpful in preparing for your exam.

contd

In addition, this chapter will guide you with **scientific data** and **uncertainties in measurement** to ensure that **valuable points** are not lost in the **practical assessments** or in the **exam**.

Our dynamic universe

The unit covers **kinematics**, **dynamics** and **space time**. General areas include:

- Motion – equations and graphs
- Forces, energy and power
- Collisions, explosions and impulse
- Gravitation
- Special relativity
- The expanding universe.

Particles and waves

The unit covers the key areas of **particles** and **waves**. General areas include:

- The standard model
- Forces on charged particles
- Nuclear reactions
- Wave particle duality
- Interference and diffraction
- Refraction of light
- Spectra.

Electricity

The unit covers **electricity**, and **electrical storage and transfer**, including:

- Monitoring and measuring a.c.
- Current, potential difference, power and resistance
- Electrical sources and internal resistance
- Capacitors
- Conductors, semiconductors and insulators
- p-n junctions.

ASSESSMENT

To gain the award of the course, you must pass all of the units as well as the course assessment. Course assessment will provide the basis for grading attainment at levels A–D. A key to **understanding assessment** is to understand that different assessments have **different types of questions** and questions with **different levels of difficulty**.

Unit assessment

You may be following a **unit by unit** approach or have a **combined assessment**. These are internally assessed. You will be expected to pass a set of **questions** and plan and carry out a **practical assessment**, submitting a report on your findings. It is useful to view the unit assessment(s) as smaller hurdles to pass before you prepare for the final exam.

Course assessment

Part 1 – question paper (scaled from 130 marks)   **100 marks**   2·5 hours

This is split into an objective test of 20 marks and questions requiring restricted and extended response and will be scaled down from 110 to 80 marks. While most marks will be for applying knowledge and understanding, your other skills will be required. You need to work on these more complex problems before your prelim and the final exams.

Part 2 – assignment   **20 marks**

This assignment requires you to apply skills, knowledge and understanding to investigate a relevant topic in physics. See pp 118–123 for more detail.

Total marks: 120 marks

A **data booklet** containing relevant data and formulae will be provided.

 ## THINGS TO DO AND THINK ABOUT

As you begin your course you would normally be expected to have attained the skills, knowledge and understanding required from National 5 Physics Course or similar. However this book will occasionally revise some relevant points to help introduce you to new material. All of which, in a nutshell, outlines your Higher Physics course. Good luck!

 **DON'T FORGET**

You can expect an '**advice to candidates**' sheet for both the questions and the practical assessment. You should check that you follow this advice carefully to ensure you pass.

 **DON'T FORGET**

The course assessments require breadth and depth from across the units and are externally assessed.

SKILLS, RESEARCH AND INVESTIGATIONS

During your course you will undertake research and practical investigations. Here we consider the **skills** you need which are relevant to physics research, important for planning and undertaking practical investigations and analysing results. You can build on knowledge and understanding learned in the course and this book.

SKILLS, KNOWLEDGE AND UNDERSTANDING

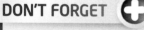

As you **study** and carry out **practical research** and **investigations** in physics, you will develop **skills** as well as **knowledge and understanding**. Here are some skills you should aim for:

- **demonstrate** knowledge and understanding of physics by making statements, describing information, providing explanations and integrating knowledge
  - **apply** physics knowledge to **new situations**, interpreting information and solving problems
    - **plan and design** experiments or practical investigations to test given hypotheses
      - **undertake experiments** competently and **safely**, recording detailed observations and collecting data
      - use a **variety** of methods to select and present information **appropriately**
- **process** information using the correct **significant figures** and **units**
- make **estimates** and **predictions**
- draw **valid** conclusions and explanations backed up by evidence or **justification**
- **evaluate** experimental procedures, **identifying** sources of error and **suggesting** improvements
- **communicate** findings or information **effectively**.

Some skills to develop: the hardest is on top

RESEARCH

When **researching** unfamiliar areas of physics, either for a **topic** you are studying, or for an **assignment**, you should start with some **written questions** which you want to find out about. You can use **printed resources**, **video** or **audio** materials, **websites** or key phrases to enter into a **search engine**. When studying the topics in this book, you will find **online** and **video** references to develop your understanding of the topic.

You may wish to pursue further links to deepen your **knowledge** or **understanding**. You will be aware that not all sources are **reliable** and so you should collect and synthesize information from different sources. Take note of the **sources used** and also note key phrases or pieces of information as you go so that you can draw up **explanations** and **conclusions** in your own report later. List any **justification** of your findings.

DON'T FORGET

As you undertake scientific enquiry, develop your knowledge and understanding, do practical work or undertake research, remember to think about the skills you should use.

INVESTIGATIONS

During your course you should gain experience of planning and carrying out practical investigative work. When you carry out a formal investigation you should use standard laboratory equipment if possible. You will see examples in graphics throughout this book. Any unfamiliar equipment required should be researched or demonstrated to you.

contd

When carrying out a formal investigation you should **report** it to a standard appropriate for a 'Higher', taking account of the following.

- **Numerical results** should be recorded in tables and graphs as appropriate. Headings and axes must be labelled and appropriate scales used.
- Lines of **best fit** to curves or straight lines should be drawn.
- Consider how **relationships** should be expressed. Straight lines could be expressed in the form $y = mx + c$ and the gradient and intercept on the y-axis used to find m and c. Or **relationships** could simply be stated as direct or indirect variation.
- **Measurements** should be **repeated** as appropriate and a mean value calculated.
- Scale-reading **uncertainties** should be estimated and expressed in absolute or percentage form.
- When measuring more than one physical quantity, the quantity with the **largest percentage uncertainty** should be identified and this can be used as an **estimate** of the percentage uncertainty in the final result.
- The **final** numerical result of an experiment should be expressed in the form: **final value ± uncertainty**.

ESTIMATES AND OPEN-ENDED QUESTIONS

Estimates

In this course you will study ranges of **scale** from sub-atomic particles to the largest expanse of the universe. You will be able to make an intelligent **estimate** of not only distances but quantities from other topics in physics. It helps to have your own mental list of familiar items to provide a mental ruler.

time	height of school	distance across our galaxy	speed of a car	speed of walking
1 hour = 3600 s	10 m	100 000 light years	30 mph = 50 kmh⁻¹	2 ms⁻¹
mass of pupil	mass of Earth	an apple weighs 1 N	power of an electric fire	room temperature
70 kg	6×10^{24} kg		2000 W	20°C

Your own list may be totally different and more extensive. The important thing is you develop this skill based on your physics experience and are able to apply it. For example, if you are asked to estimate a value in an investigation or in an exam, your estimate can allow you to make further **predictions**. If you are asked to estimate the speed of a car after 5 s from rest, you can then also make calculations of the car's acceleration. The exam board will allow a range of reasonable estimates and credit the further workings.

Do watch out for **reasonable orders of magnitude**. For example, an aircraft runway may be 2×10^3 m long but not 2×10^1 m.

Open-ended questions

An open ended question is typically of the form *Use your knowledge of physics to explain … .*

You should make a statement of **principle(s)** involved, and/or a relationship or **equation**, and apply these to respond to the problem/situation. Develop and justify your argument. You will gain more marks for **breadth** and **depth** of understanding. Be aware you may have to correct a false statement. **Explain** how the physics applies.

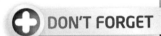

DON'T FORGET

The data sheet in the exam provides data you may require and you should use these values.

ONLINE

Find some practical topics at www.brightredbooks.net

VIDEO LINK

Check out the Richard Feynman physics investigations at www.brightredbooks.net

THINGS TO DO AND THINK ABOUT

Use this chapter to help you develop your skills in more detail and to a Higher standard throughout your course.

SCIENTIFIC DATA

Learn how the units of physics are based on an international system, learn prefixes, ensure you are able to use scientific notation, and understand the correct use of significant figures. You will need these through your Higher course.

ONLINE

Read more at www. brightredbooks.net

SI UNITS

The Higher Physics course uses the Système International d'Unités (SI) or International System of units. This consists of a small number of SI base units and a bigger number of derived units. You will find a full list of the physical quantities and their units on the inside back cover.

SI Base units

UNIT	SYMBOL	QUANTITY
metre	m	length
kilogram	kg	mass
second	s	time
ampere	A	electrical current
kelvin	K	temperature

Most symbols for units are written in lower case unless named after a person. Symbols are not pluralised, and they do not close with a full stop. For large numbers, you should group numbers in threes with a space between each group (instead of a comma) and to put a space before the unit. For example: $1\,957\,984\,\text{ms}^{-1}$.

Prefixes and powers of 10

In large or small numbers, prefixes before the unit indicate a multiple of the unit in powers of 10.

FACTOR		NAME	SYMBOL
10^{12}	1 000 000 000 000	tera	T
10^{9}	1 000 000 000	giga	G
10^{6}	1 000 000	mega	M
10^{3}	1000	kilo	k
10^{-2}	0.01	centi	c
10^{-3}	0.001	milli	m
10^{-6}	0.000 001	micro	μ
10^{-9}	0.000 000 001	nano	n
10^{-12}	0.000 000 000 001	pico	p

Powers of 10

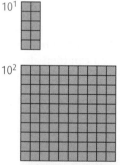

10^{0}
10^{1}
10^{2}
10^{3}

Example:

$14\,000\,\text{m} = 14\,\text{km}$ $4\,700\,000\,\Omega = 4\cdot7\,\text{M}\Omega$
$0\cdot000\,008\,\text{s} = 8\,\mu\text{s}$
$15\,\text{THz} = 15\,000\,000\,000\,000\,\text{Hz}$
$450\,\text{nm} = 0\cdot000\,000\,450\,\text{m}$

DON'T FORGET

If you don't watch out for the prefixes in a question, you'll get the wrong answer.

SCIENTIFIC NOTATION

Scientific notation is used to write large and small numbers and helps avoid making mistakes counting zeroes. For example,

- the Earth's mass is about $5\,973\,600\,000\,000\,000\,000\,000\,000\,\text{kg}$. It is much easier to write $5{\cdot}9736 \times 10^{24}\,\text{kg}$.

- the mass of a proton is $0{\cdot}000\,000\,000\,000\,000\,000\,000\,000\,001\,673\,\text{kg}$. It is much easier to write $1{\cdot}673 \times 10^{-27}\,\text{kg}$.

Make sure you know how to use standard notation on your calculator.

Questions

Multiplication: $(3 \times 10^8) \times (2 \times 10^5) = (3 \times 2) \times 10^{(8+5)} = 6 \times 10^{13}$

Division: $\dfrac{3 \times 10^8}{2 \times 10^5} = \left(\dfrac{3}{2}\right) \times 10^{(8-5)} = 1{\cdot}5 \times 10^3$

DON'T FORGET

Scientific notation has a number of useful properties and is commonly used in calculators and by scientists, mathematicians and engineers.

SIGNIFICANT FIGURES

A calculation using measurements cannot have more accuracy than the figures of the measurements.

$1\,\text{m}$ divided by $3\,\text{s}$ cannot be $0{\cdot}333\,333\,3\,\text{ms}^{-1}$ even if your calculator says so.

As there is only one significant figure in each measurement, the answer should also have only one significant figure.

$$v = \frac{s}{t} = \frac{1}{3} = 0{\cdot}3\,\text{ms}^{-1}$$

$3\,\text{s}$, $3{\cdot}0\,\text{s}$, $3{\cdot}00\,\text{s}$, and $3{\cdot}000\,\text{s}$ all have the same value, but if a figure is shown as $3{\cdot}000\,\text{s}$, we assume it has been measured more precisely. $1{\cdot}00\,\text{m}$ is more precise than $1\,\text{m}$.

$$v = \frac{s}{t} = \frac{1{\cdot}00}{3{\cdot}000} = 0{\cdot}333\,\text{ms}^{-1}$$

If your calculator produces too many significant figures, you can carry these through until the end of your working, but you must then **remember to round up or down** as appropriate.

You should **round to the smallest number of significant figures** in the measurements.

$$\frac{24{\cdot}92}{3{\cdot}64} = 6{\cdot}846\,153\,8$$

$6{\cdot}846\,153\,8$ rounded to 3 significant figures is $6{\cdot}85$.

Examples:

30 has 1 significant figure. 30·0 has 3 significant figures.
0·007 020 0 has 5 significant figures. $5{\cdot}40 \times 10^3$ has 3 significant figures.
5400 has 2 significant figures.
6·846 153 8 rounded to 3 significant figures is 6·8.

DON'T FORGET

Always check that your answer to a calculation looks reasonable and possible. Scientific notation is the easiest way for large numbers.

DON'T FORGET

Round at the end of a calculation.

VIDEO LINK

Check out the clips on prefixes to standard form and significant figures at www.brightredbooks.net

THINGS TO DO AND THINK ABOUT

1 Make sure you add the correct unit to an answer to gain that mark.
2 Missing a prefix in a question puts your answer out by powers of 10.
3 You can do a rough check on answers just using the powers of 10 in scientific notation.
4 Keep an additional significant figure in intermediate values to calculations.
5 Zeros are important.
6 Use the data sheet values in calculations.

ONLINE

Head to www.brightredbooks.net for more on scientific notation and significant figures.

UNCERTAINTY IN MEASUREMENT

When a physical quantity is measured, it will always be liable to an **uncertainty**. There are different **types of uncertainty** which can all contribute to uncertainty in measurements.

The smallest division is 1 V, half this = 0·5 V.
The reading is:
3 ± 0·5 V

The smallest division is 0·1 A, half this = 0·05 A.
The reading is:
0·35 ± 0·05 A

DON'T FORGET ➕

How well a scale can be read is only one source of uncertainty.

READING UNCERTAINTIES

Scale-reading uncertainties indicate the accuracy to which an instrument scale can be read. Scale-reading uncertainties exist in both analogue and digital scales.

Analogue scales

Reading uncertainty for an analogue scale is **±0·5 of the smallest division**.
This allows us to read to the nearest half division.

Example:
How well can you read these scales?

An exception is made for wide divisions where a more reasonable estimate would be ±0·2 of the smallest division.
This allows us to read to the nearest one fifth of a wide division.
The smallest division is 1 cm, one fifth of this = 0·2 cm

Example:
How well can you read this scale?

Solution:
The reading is: 12·6 ± 0·2 cm

Digital scales

Reading uncertainty for a digital scale is **±1 on the least significant digit**.
This allows us to read the least significant digit.

Example:
What is the reading uncertainty in this digital stop-watch?

Solution:
The least significant digit is 0·01 s.
The reading is: 3·26 ± 0·01 s

RANDOM UNCERTAINTIES

When a measurement of a physical quantity is repeated, we usually see different readings. There is equal probability that each measurement made is higher or lower than the 'true' value. Repeated measurements of a physical value are desirable. The mean of repeated measurements is the best estimate of a 'true' value being measured.
The mean is found from the sum of the measurements divided by the number of measurements.

$$\text{mean} = \frac{\text{sum of values}}{\text{number of values}}$$

Example:
The time taken for a falling ball to hit the ground is measured five times and the following results obtained:
t = 1·48, 1·32, 1·38, 1·37, 1·40 s.

Solution:
The mean is calculated.
$$\text{mean} = \frac{6·95}{5} = 1·39\,s$$

contd

Repeating measurements reduces uncertainty. The greater the number of repeats, the smaller the uncertainty.

The random uncertainty is estimated by dividing the range of the measurements by the number of measurements.

$$\text{random uncertainty} = \frac{\text{maximum value} - \text{minimum value}}{\text{number of values}}$$

Example:

The uncertainty for the falling ball in the previous examples is:

uncertainty $= \frac{1\cdot48 - 1\cdot32}{5} = \frac{0\cdot16}{5} = 0\cdot032 = 0\cdot03$

Experimental measurements should be expressed in the form:
value ± uncertainty
So, the time taken for the falling ball should be shown as $t = 1\cdot39 \pm 0\cdot03\,$s.

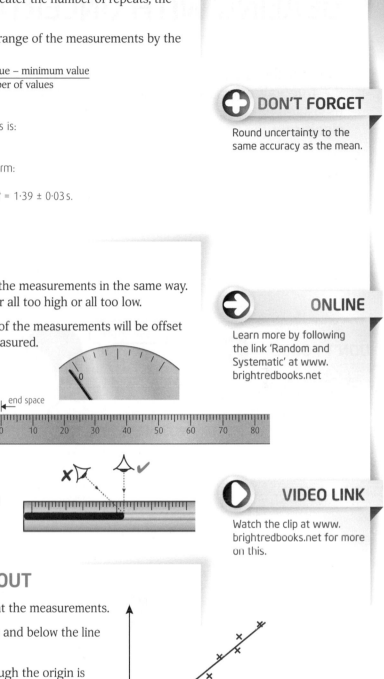

DON'T FORGET

Round uncertainty to the same accuracy as the mean.

SYSTEMATIC UNCERTAINTIES

A systematic uncertainty is an error which affects all the measurements in the same way. A systematic uncertainty will make the readings either all too high or all too low.

Where a systematic effect is present, the mean value of the measurements will be offset from the true value of the physical quantity being measured.

Faults in the apparatus

A meter may not be set to its zero value properly.

A ruler may be used without allowing for the end space.

Experimental technique

A scale may always be read from one side rather than directly in front.

ONLINE

Learn more by following the link 'Random and Systematic' at www. brightredbooks.net

VIDEO LINK

Watch the clip at www. brightredbooks.net for more on this.

THINGS TO DO AND THINK ABOUT

1   To increase the accuracy of a measurement, repeat the measurements.

2   The crosses or points on this graph lie both above and below the line of best fit indicating random uncertainties.

3   On a graph: a straight line which nearly goes through the origin is likely to have had measurements offset in one way. This could be a systematic uncertainty due to apparatus.

4   What is the difference between random uncertainties and systematic effects?

Straight line, nearly through origin.

Possible systematic uncertainty?

DEALING WITH UNCERTAINTIES

The **measurement** of any physical quantity is liable to **uncertainty**, so we have to **estimate** the **true value** (e.g. by calculating the mean) and then **estimate** the **uncertainty** in the measurement (e.g. by using the random uncertainty equation). We now learn to consider how **significant** the uncertainty is and how to **combine** several uncertainties.

ABSOLUTE AND PERCENTAGE UNCERTAINTIES

Absolute form

The uncertainty can be expressed with the same units as the value measured. This is known as the absolute uncertainty. The results are written as shown:

estimated value $\longleftarrow$ $\begin{array}{l} 2{\cdot}78 \pm 0{\cdot}03 \text{ mA} \\ 5{\cdot}6 \pm 0{\cdot}1 \text{ } \mu\text{V} \\ 15{\cdot}23 \pm 0{\cdot}50 \text{ s} \end{array}$ $\longrightarrow$ absolute uncertainty and unit

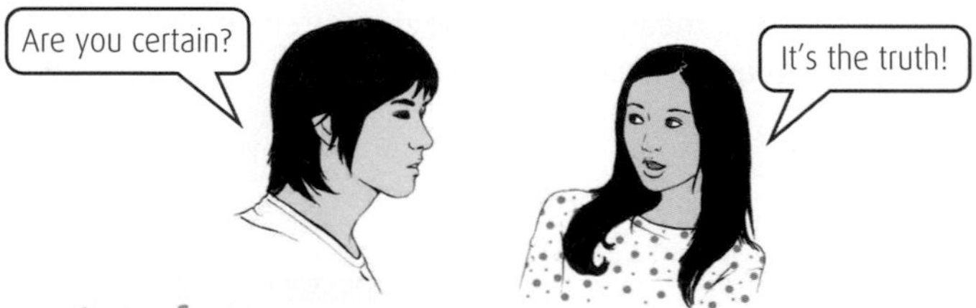

Are you certain?

It's the truth!

Percentage form

Uncertainty can be expressed in percentage form. Percentage uncertainty allows us to indicate how precise a value is. The percentage uncertainty has to be calculated:

% uncertainty = $\dfrac{\text{absolute uncertainty}}{\text{measurement}} \times 100$

Check you can obtain these percentage uncertainties for the absolute uncertainties shown above:

2·78 mA, 1%   5·6 μV, 2%   15·23 s, 3%

> **Example:**

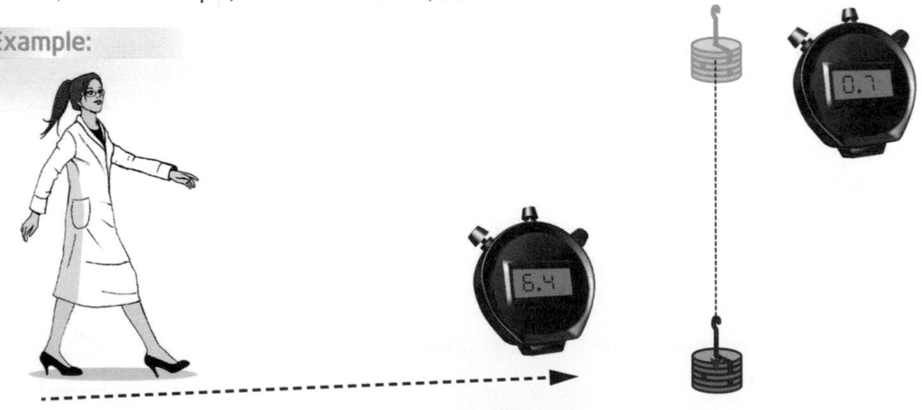

A young physicist walks across the room. She has measured her time to do this with a stop-watch. She records a time of 6·4 s. She then records the time it takes a mass to fall as 0·7 s. She estimates her uncertainty in using the stop-watch to be 0·2 s.
Express her results in absolute form and calculate the percentage uncertainty for each.

> **Solution:**

Physicist's walk: $t = 6{\cdot}4 \pm 0.2$ s % uncertainty = $\dfrac{0{\cdot}2}{6{\cdot}4} \times 100 = 3\%$

Falling mass: $t = 0{\cdot}7 \pm 0.2$ s % uncertainty = $\dfrac{0{\cdot}2}{0{\cdot}7} \times 100 = 29\%$

We can see that the same absolute uncertainty has a greater percentage effect when timing a short time interval than when timing over a long time.

COMBINING UNCERTAINTIES

In doing an experiment, several variables can be measured, and then the final value can be calculated.

The final numerical result of an experiment should be given as **final value ± uncertainty**.

The final value should be calculated from the measurements using an equation from the data sheet in the normal way.

The final or overall uncertainty needs to be calculated.
- Calculate the percentage uncertainty for each measurement.
- The largest percentage uncertainty is the most significant.
- The largest percentage uncertainty is a good estimate of the percentage uncertainty in the final numerical result of the experiment.
- Use this percentage uncertainty to calculate the absolute uncertainty from the final value.

DON'T FORGET

Select the largest percentage uncertainty as the one to use.

Example:

A physicist is asked to find out how much energy is used when a kettle is boiled. He takes the following measurements.

Voltage of supply	$V = 225 \pm 25\,V$
Current drawn	$I = 9 \cdot 5 \pm 0 \cdot 5\,A$
Time taken	$t = 125 \pm 5\,s$

Solution:

Energy $E = ItV = 9 \cdot 5 \times 125 \times 225 = 267\,188\,J$
Percentage uncertainties:

$V \quad \dfrac{25}{225} \times 100 = 11\%$

$I \quad \dfrac{0 \cdot 5}{9 \cdot 5} \times 100 = 5\%$

$t \quad \dfrac{5}{125} \times 100 = 4\%$

The largest percentage uncertainty is 11%.
The final uncertainty is 11% of $267\,188\,J = 29\,391\,J$
Remember to round to the fewest significant figures.
The final result is:
Energy $E = 0 \cdot 27 \pm 0 \cdot 03\,MJ$

kettle

voltmeter

ammeter

stop-watch

DON'T FORGET

It is not enough to know a value or measurement, we have to know how well we know it.

Improvements

We should consider how to improve experiments. A common solution is to **repeat readings** of measurements and use meters with **smaller divisions**. We need to identify the **biggest cause** of uncertainty.

In the experiment above the voltage has the biggest uncertainty. Improving our measurements of current or time will not make any significant difference unless we first reduce the uncertainty for the voltage. The voltmeter must have very large divisions to have a ±25 V uncertainty. If we use a meter which reads to the nearest 1 volt, the uncertainty is reduced to ±1 V or ±0·4%. Take a number of readings and calculate the mean. Then we can consider current and time.

THINGS TO DO AND THINK ABOUT

1 You are asked to time the period of a pendulum. How would you reduce the uncertainty in the measurement of the period (time for one swing)?

Do not try to time one swing. The time will be very short. The percentage uncertainty will be high. The absolute uncertainty is the same for one swing or 10 swings. Time a number of periods (say 10) and divide the result by the number of swings. The absolute uncertainty is then also divided by 10 when you calculate one period.

2 Whenever you are asked to do a formal report of an **investigation**, you should think of all the uncertainties in each measurement and remember how we combine them if you calculate a value.

SCALARS AND VECTORS

In this section, we will consider the idea of motion, how to combine scalar and vector quantities, and use graphical and mathematical methods when dealing with scalar and vector quantities. Through this course you will meet both scalar and vector quantities.

Know the resultant direction from start to finish.

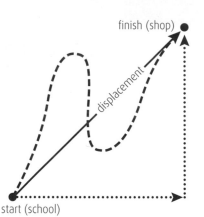

Different paths, same displacement

MOTION

A body or an object is said to be in motion when it changes its position. We can analyse an object's displacement, velocity and acceleration during its change of position. These are **vector** quantities so both **magnitude** (size) and **direction** must be stated for each. A **scalar** quantity has no direction.

SCALAR (MAGNITUDE ONLY)	VECTOR (MAGNITUDE AND DIRECTION)
Time	
Energy	
Mass	
Distance	Displacement
Speed	Velocity
	Acceleration
	Momentum
	Force
	Impulse

Displacement

Displacement is the distance travelled in a certain direction from the start point to the finish point in a straight line.

Velocity

Velocity is the rate of change of displacement. Velocity is the displacement covered in a certain time. Direction has to be stated.

Acceleration

Acceleration is the rate of change of velocity. Acceleration is the change of velocity in a certain time. Acceleration takes place in a certain direction.

DON'T FORGET ✚

You may have met scalars and vectors in National 5 Physics. Make sure you revise these quantities here as they form an important part of the Higher course.

DON'T FORGET ✚

When drawing your scale diagram keep it large to reduce the uncertainty in the answer.

SCALE DRAWINGS

Vector quantities can be represented with a line drawn to a scale. An arrowhead must be added to show which direction the line represents. Remember these rules to add vector quantities together.

1 Choose and write down a scale.

2 Add vectors drawn 'head to tail'.

3 Draw the resultant from 'start to finish'.

4 Measure both the magnitude and the direction.

An arrow has magnitude and direction.

MATHEMATICAL METHODS

Mathematical methods can be used as an alternative to scale diagrams when adding vector quantities together or resolving a vector into components.

You must remember to calculate both magnitude and direction.

Right-angled triangles

Use Pythagoras' Theorem to calculate the resultant magnitude of two vectors at right angles:

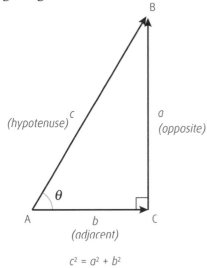

$$c^2 = a^2 + b^2$$

Alternatively, you can use trigonometry to calculate the direction:

$$\sin\theta = \frac{\text{opposite}}{\text{hypotenuse}}$$

$$\cos\theta = \frac{\text{adjacent}}{\text{hypotenuse}}$$

$$\tan\theta = \frac{\text{opposite}}{\text{adjacent}}$$

Acute or obtuse angled triangles

Use the sine rule to calculate the size of an angle or length of a side when you have two sides and an angle or two angles and a side.

$$\frac{a}{\sin A} = \frac{b}{\sin B} = \frac{c}{\sin C}$$

Use the cosine rule to calculate the length of a side or size of an angle when you have two sides and an angle or three sides.
$$a^2 = b^2 + c^2 - 2bc\cos A$$

THINGS TO DO AND THINK ABOUT

When an object falls to the ground its velocity is said to be increasing and it has an acceleration which is vertically downwards towards the centre of the Earth. An object in orbit is usually said to have a constant orbital speed but gravity is changing its direction to keep it in orbit. This object is also said to be accelerating towards the Earth. How can an object with constant speed also have acceleration?

Orbits of satellites accelerating

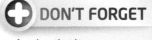

ONLINE

For more on vectors, head to www.brightredbooks.net

ONLINE TEST

Test your knowledge of this topic at www.brightredbooks.net

VIDEO LINK

Head to www.brightredbooks.net for an explanation of the 'tip-to-toe' method for adding vectors.

DON'T FORGET

Acceleration is a vector quantity.

DISPLACEMENT AND VELOCITY

In this section we will consider the vector nature of displacement and velocity. We will also learn how to use vector rules to add these quantities.

DISTANCE AND DISPLACEMENT

A body in motion will often change its direction. The distance travelled from start to finish can vary with the routes taken but displacement is defined as the straight-line distance and direction from start to finish.

The symbol for distance is d. Distance is defined by a number and its unit, the metre, m. For example, the length of the school laboratory is 8 m.

Displacement is a vector quantity. For example, an explorer walks 20 km due north from base camp. The symbol for displacement is s.

Example:

1. A pupil travels east 10 m along the school corridor but is sent back 7 m for running. Find his displacement.
2. A car drives 4 km north then drives 3 km west. What is the resultant displacement?
3. A runner goes round a running track three times. If the track has a length of 400 m, what is the runner's distance travelled and displacement?

Solution:

1. distance travelled d = 10 + 7 = **17 m**
   displacement s = 10 + (–7) = **3 m east**.

If the pupil does this again, the displacement will still only be **6 m east**.

2. Using Pythagoras' theorem and trigonometry.
   $x^2 = 4^2 + 3^2$
   $x = 5$ km
   $\tan\theta = \frac{3}{4}$
   $\theta = 37°$
   displacement, s = **5 km at 37° to the west of north**
   Use a ruler and protractor to check this by scale measurement.

3.

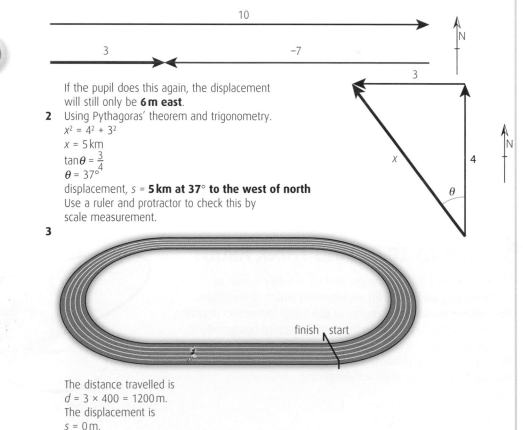

The distance travelled is
$d = 3 \times 400 = 1200$ m.
The displacement is
$s = 0$ m.

> **DON'T FORGET**
>
> The magnitude of the displacement is different from the distance when an object changes direction.

> **DON'T FORGET**
>
> If we ignore direction when calculating vector quantities, we will usually get the wrong result.

SPEED AND VELOCITY

Speed is a **scalar** quantity. Speed is the distance travelled in unit time.

$$\text{speed} = \frac{\text{distance}}{\text{time}}$$

Speed only has **magnitude**, no direction, so a description such as 'The car was going faster than $30\,\text{ms}^{-1}$' refers to a speed.

Velocity is a **vector** quantity. Velocity is the displacement per unit time.

$$\text{velocity} = \frac{\text{displacement}}{\text{time}} \text{ or } v = \frac{s}{t}$$

Velocity has **magnitude** and direction, so a description such as 'The car was travelling at $30\,\text{ms}^{-1}$ **from Glasgow to Edinburgh**' refers to a velocity v.

Example:

A sailor sets his boat on a heading of north at $5\,\text{ms}^{-1}$ through the sea. The tide is moving at $2\,\text{ms}^{-1}$ in a south-east direction.
Find the boat's resultant velocity.

Solution:

Scale: $1\,\text{cm} = 1\,\text{ms}^{-1}$

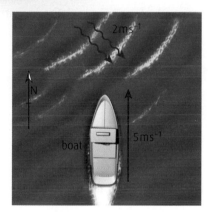

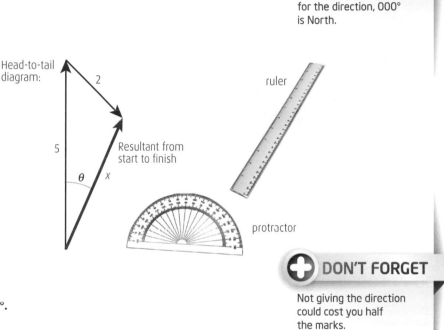

Measure the resultant $x = 3.9\,\text{cm}$
Measure the angle $\theta = 22°$
The resultant velocity is $\boldsymbol{v = 3.9\,\text{ms}^{-1}}$ **at 022°.**

THINGS TO DO AND THINK ABOUT

The use of s for displacement is from the Latin word for space *spatium*, first used in Galileo's *Discourses on Two New Sciences* in 1640.

1   Use your knowledge of vectors to explain to a friend how your velocity changes when
    - you walk on a conveyor belt
    - a plane heads into a crosswind
    - you sail against the tide.

2   A destination is said to be $48\,\text{km}$ away. Explain as if to a passenger why your journey might take 6 hours even though the car you are driving will be able to maintain an average speed of $12\,\text{kmh}^{-1}$.

3   Compare distance and displacement in this image.

ACCELERATION AND FORCE

In this section we learn about the use of light gates to measure velocity and acceleration. We will also relate force and acceleration as vector quantities.

MEASURING VELOCITY USING LIGHT GATES

Single light gate connected to a motion computer

A good estimate of the magnitude of an instantaneous velocity can be obtained by measuring a small change in displacement over a small change in time. The displacement of a card mask through a light gate can be timed and the software will calculate the velocity at that point.

$$v = \frac{\Delta s}{\Delta t}$$

(Δ = 'change in')

Δs

motion computer

light gate

- Use a ruler to measure the length of card. Enter this value into the software.

- The light gate and motion computer measure the short time interval for the card mask to cut the light beam.

- The software will then use the velocity equation to calculate the magnitude of the velocity.

MEASURING ACCELERATION USING LIGHT GATES

Acceleration is the change in velocity per unit time.

$$a = \frac{\Delta v}{\Delta t} \quad \text{or} \quad a = \frac{v - u}{t}$$

(u = first velocity, v = second velocity, t is the time between.)

Using two light gates and a card mask

The magnitude of the average acceleration on a slope can be obtained by measuring a change in velocity over a change in time.

- You will need to **measure** and **enter** the **length of card** into the software.

- Run the vehicle down the slope so that the card cuts both light beams. The motion computer measures these **two individual times** for the beam to be cut.

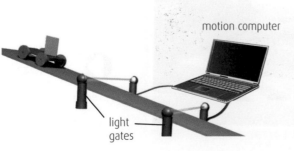

motion computer

light gates

- Now the computer can **calculate** the two velocities at the beams using the **velocity equation**.

- To calculate the change in velocity, you will need to measure and enter the time interval t it takes for the vehicle to pass the light gates. This can be done with a **stop-watch**, or the **computer** may have been programmed to capture this time also.

- Now the computer can **calculate** the **acceleration** on the slope using the **acceleration equation**.

contd

> **DON'T FORGET**
>
> The average acceleration between the light gates is the same as the acceleration at a certain point between these two light gates.

Using a single light gate and a double card mask

A good estimate of the magnitude of the acceleration at a point can be obtained by measuring a small change in velocity over a small change in time.

$$a = \frac{\Delta v}{\Delta t} \quad \text{or} \quad a = \frac{v - u}{t}$$

- The light gate and motion computer measure the times for card 1, card 2 and the time interval between as the vehicle passes through the light gate.

- The velocities u and v are calculated using **the velocity equation**.

- The **acceleration equation** can then be used to calculate the **magnitude** of the acceleration.

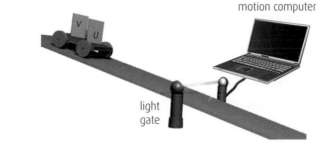

motion computer

light gate

ACCELERATION AND FORCE

Acceleration and force are both vector quantities. Acceleration is directly proportional to force.

An object moving down a slope is subject to a component of gravity to provide the acceleration down the slope. The force and the acceleration are both in the same direction down the slope.

An object falling to the ground is also subject to gravity. The downward force of gravity produces acceleration downwards.

The acceleration of any object will be in the same direction as the unbalanced force acting on it.

F down

a down

THINGS TO DO AND THINK ABOUT

Using the apparatus described above, measure the value of acceleration with different angles of slope. The force of gravity is vertically downwards. Why are the acceleration and the accelerating force on a slope both in the direction of the slope? What can make their values change? This will be studied in detail shortly.

The force and acceleration of the Moon, which is in orbit around the Earth, are towards the Earth even though the Moon is travelling 'sideways'. Why?

EQUATIONS OF MOTION

For an object with constant acceleration in a straight line, we learn how to derive alternative equations or relationships that use the *suvat* quantities. You will successfully apply these equations to objects in motion.

s = displacement u = initial velocity v = final velocity a = acceleration t = time

Acceleration is the **rate of change of velocity**. This applies whether an object is increasing or decreasing its velocity. If the velocity of an object is decreasing, the acceleration may have a negative value.

DERIVING EQUATIONS OF MOTION

Equation 1: Rearranging the acceleration equation

We can rearrange the familiar equation which defines acceleration to create our first new equation for acceleration a.

$$a = \frac{v - u}{t}$$

This can be rearranged to find the final velocity v

$$v - u = at$$

so final velocity, $v = u + at$ Equation 1

In words, this expression says that the final velocity of an object depends on its initial velocity plus the fact that it accelerates for a time. Also note that the product, at, represents the change of velocity.

This equation links acceleration, initial and final velocities, and time but not displacement.

Equation 2: Derived from a velocity-time graph

An object is accelerating from initial velocity $\boldsymbol{u}$ to a final velocity $\boldsymbol{v}$ in time $\boldsymbol{t}$.

Consider the graph of this motion. (Note this sketch graph has its axes fully labelled.)

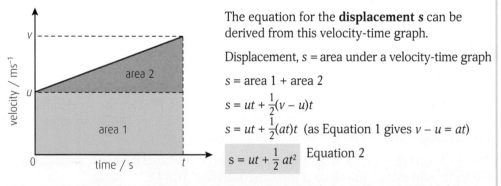

The equation for the **displacement s** can be derived from this velocity-time graph.

Displacement, s = area under a velocity-time graph

s = area 1 + area 2

$s = ut + \frac{1}{2}(v - u)t$

$s = ut + \frac{1}{2}(at)t$ (as Equation 1 gives $v - u = at$)

$s = ut + \frac{1}{2}at^2$ Equation 2

This equation links acceleration, initial velocity, displacement and time but not the final velocity. Note this equation simplifies to the basic equation $s = ut$ only when $a = 0$.

By comparing our new equation with the graph we can also note

area 1 displacement = ut (displacement without acceleration)

and   area 2 displacement = $\frac{1}{2}at^2$ (additional displacement due to the acceleration)

contd

Equation 3: Combining Equations 1 and 2

Equations 1 and 2 can be combined into a third useful equation.

From Equation 1

$$v = u + at$$

square both sides $\quad v^2 = (u + at)^2$

open the brackets $\quad v^2 = u^2 + 2uat + a^2t^2$

use common factor $2a$ $\quad v^2 = u^2 + 2a(ut + \frac{1}{2}at^2)$

From Equation 2

$$(ut + \frac{1}{2}at^2) = s$$

substituting $\quad v^2 = u^2 + 2as \quad$ Equation 3

This equation links acceleration, initial and final velocities, and displacement but not time.

Equation 4 from average velocity

Displacement of an object with an average velocity $\quad \bar{v} \qquad\qquad s = \bar{v}t$

The average velocity is halfway between initial and final velocities $\quad \bar{v} = \frac{1}{2}(v + u)$

Combining these gives

$$s = \frac{1}{2}(v + u)t \quad \text{Equation 4}$$

This equation links initial and final velocities, displacement and time but not acceleration. The equation is useful when you do not have the acceleration value but you know the object has constant acceleration.

Equations summarised

Identify the equation that fits the problem best.

$v = u + at \quad$ Equation 1

$s = ut + \frac{1}{2}at^2 \quad$ Equation 2

$v^2 = u^2 + 2as \quad$ Equation 3

$s = \frac{1}{2}(v + u)t \quad$ Equation 4

EQUATION	s	u	v	a	t
1	0	✓	✓	✓	✓
2	✓	✓	0	✓	✓
3	✓	✓	✓	✓	0
4	✓	✓	✓	0	✓

THINGS TO DO AND THINK ABOUT

s, u, v and a are vectors and so direction must be considered. t is treated as a scalar.

Although you may not have to derive these equations in an exam, you should be able to follow these derivations. This is a skill in applying your mathematical methods which a Higher level student should develop.

Before we had all these equations we could only measure or calculate acceleration using the standard acceleration equation. Now, for example, we can calculate acceleration without measuring time, see equation 3.

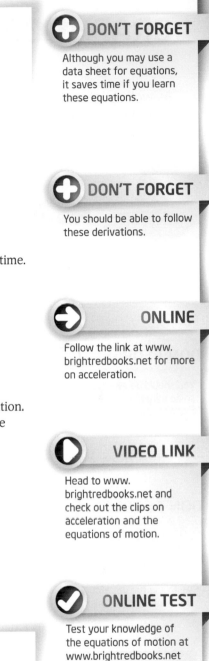

DON'T FORGET

Although you may use a data sheet for equations, it saves time if you learn these equations.

DON'T FORGET

You should be able to follow these derivations.

ONLINE

Follow the link at www. brightredbooks.net for more on acceleration.

VIDEO LINK

Head to www. brightredbooks.net and check out the clips on acceleration and the equations of motion.

ONLINE TEST

Test your knowledge of the equations of motion at www.brightredbooks.net

GRAPHS OF MOTION

As well as learning how to plot an accurate graph, you also need to learn how to interpret graphs of motion and be able to visualise the shape of a graph for different motions. You will need to compare displacement-time, velocity-time and acceleration-time graphs for the same motion.

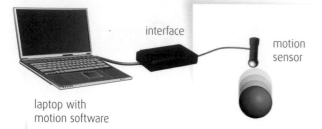

interface

motion sensor

laptop with motion software

MOTION SENSORS

A motion sensor sends out pulses of ultrasound. These are reflected from a moving object and the length of time from transmission to reception is used to calculate the displacement, velocity and acceleration of the object with time. The results can be displayed in tables or graphs.

DON'T FORGET ➕

You should be aware of some equations for velocity and acceleration that the motion software can use.

DON'T FORGET ➕

Cover this page with a sheet of paper, move it down slowly, then try to predict the graphs which will follow the description of the motion. This will help you learn.

OBJECTS IN MOTION

Train on a straight track

Consider a train moving at a **constant velocity** along a straight track. A velocity-time graph may be familiar but what will a displacement-time and acceleration-time graph look like?

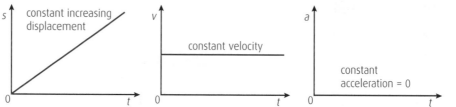

The velocity-time graph shows a constant value of velocity. The displacement-time graph shows that the train is moving away from a start point at a constant rate and the acceleration-time graph shows a constant zero value of acceleration.

Lab trolley on a straight slope

Consider a lab trolley **accelerating** down a raised straight slope in the school lab. Can you predict what the displacement-time, velocity-time and acceleration-time graphs will look like? Can you label each graph with a description of the motion?

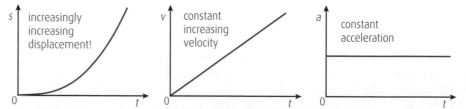

The velocity is increasing constantly so the velocity-time line is also increasing constantly. The distance travelled from the start is increasing at an increasing rate so the displacement-time graph curves up. The trolley is accelerating but the acceleration value is constant so the acceleration-time graph maintains a constant horizontal line.

contd

Car on a short journey

A car accelerates from rest, then maintains a constant velocity before decelerating to rest. Draw the velocity-time and acceleration-time graphs for this motion.

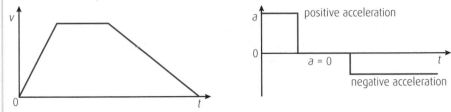

You can use values on the velocity-time graph to calculate values for the acceleration-time graph.

Ball thrown vertically up

Graphs can show the motion of a ball thrown vertically upwards, from the moment it leaves a person's hand till it returns to that point.

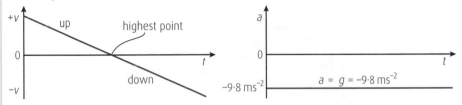

The velocity leaves the hand with a high positive (up) velocity, reduces to zero at the highest point, then increases to a high value again but in the opposite (down) direction. The acceleration always has a value of $-9.8\,\mathrm{ms^{-2}}$ in the downward direction as the force of gravity is always acting on the ball in the downward direction.

Bouncing ball

A ball is dropped from a hand to the floor where it bounces twice before being caught again.

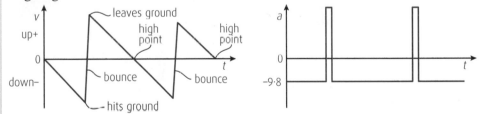

The value of the acceleration is equal to the value of acceleration due to gravity when the ball is in the air, but the value of acceleration while the ball is on the ground is usually much higher as the velocity changes direction in a very short time when it bounces. The gradients of the velocity-time graph give the values of these accelerations.

 **DON'T FORGET**

You should be able to draw an accurate a-t graph, given a v-t graph.

 **ONLINE**

Check out the link 'Bouncing ball physics' at www.brightredbooks.net

DON'T FORGET

You need to learn the rules for vector graphs as displacement, velocity and acceleration are all vectors. You should now be able to explain the physics of motion using words, diagrams, equations and graphs.

 **VIDEO LINK**

Explore this topic further by watching the clips at www.brightredbooks.net

ONLINE TEST

Revise your knowledge of graphs of motion at www.brightredbooks.net

🗨 THINGS TO DO AND THINK ABOUT

What is the significance of the gradient of the graphs and the area under the graphs?

The gradient of a **displacement-time** graph = **velocity**.

$\left(\text{velocity} = \dfrac{\text{change of displacement}}{\text{time}}\right)$

The gradient of a **velocity-time** graph = **acceleration**.

$\left(\text{acceleration} = \dfrac{\text{change of velocity}}{\text{time}}\right)$

The area under a **velocity-time** graph = **displacement**. (displacement = velocity × time)

The area under an **acceleration-time** graph = Δ velocity. (Δ **velocity** = acceleration × time)

The velocity-time graph of a ball thrown up changes from the positive to the negative, up to down. What would a speed-time graph of this motion look like?

CALCULATIONS

For an object in motion which has constant acceleration in a straight line, we learn how to calculate a physical quantity such as s = displacement, u = initial velocity, v = final velocity, a = acceleration, or t = time.

CALCULATING MOTION ON A SLOPE

To help choose the correct equation of motion, write down the *suvat* variables at the side of the page.

1   A vehicle runs down a straight track, from rest, a distance of 0·5 m in a time of 0·6 s. Calculate the vehicle's acceleration.

$s = 0\cdot5\,\text{m}$
$u = 0\,\text{ms}^{-1}$
$v = ?$
$a = ?$
$t = 0\cdot6\,\text{s}$

$s = ut + \frac{1}{2}at^2$

$0\cdot5 = 0 + \frac{1}{2}a \times 0\cdot6^2$

$a = 2\cdot8\,\text{ms}^{-2}$
acceleration, $a = 2\cdot8\,\text{ms}^{-2}$

2   A vehicle runs down a track, from rest, a distance of 0·8 m before a computer timer measures its velocity to be 1·55 ms⁻¹. Calculate the vehicle's acceleration.

$s = 0\cdot8\,\text{m}$
$u = 0\,\text{ms}^{-1}$
$v = 1\cdot55\,\text{ms}^{-1}$
$a = ?$
$t = ?$

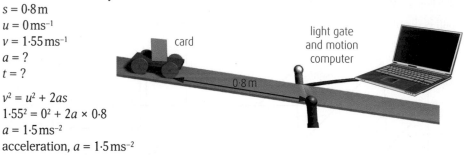

$v^2 = u^2 + 2as$
$1\cdot55^2 = 0^2 + 2a \times 0\cdot8$
$a = 1\cdot5\,\text{ms}^{-2}$
acceleration, $a = 1\cdot5\,\text{ms}^{-2}$

3   A plane, taxiing down the runway at 20 ms⁻¹, then accelerates at 3 ms⁻² for 20 s before take-off. What was the plane's take-off velocity, and what was its distance travelled while accelerating?

$s = ?$
$u = 20\,\text{ms}^{-1}$
$v = ?$
$a = 3\,\text{ms}^{-2}$
$t = 20\,\text{s}$

$v = u + at$
$v = 20 + 3 \times 20$
$v = 80\,\text{ms}^{-1}$

$s = ut + \frac{1}{2}at^2$

$s = (20 \times 20) + (\frac{1}{2} \times 3 \times 20^2)$

$s = 1000\,\text{m}$

CALCULATIONS ON VERTICAL MOTION

4   A ball is thrown upwards with an initial velocity of 6 ms⁻¹. Find the height reached and how long it takes to reach the high point.

In this question you will need to assume that the acceleration due to gravity is $-9.8\,\text{ms}^{-2}$ and the final velocity reached is $0\,\text{ms}^{-1}$.

$s = ?$
$u = 6\,\text{ms}^{-1}$
$v = 0\,\text{ms}^{-1}$
$a = -9.8\,\text{ms}^{-2}$
$t = ?$

To find the time taken
$v = u + at$
$0 = 6 + (-9.8 \times t)$
time to top, $t = 0.61\,\text{s}$

To find the height reached
$v^2 = u^2 + 2as$
$0^2 = 6^2 + 2 \times (-9.8) \times s$
$s = 1.8\,\text{m}$
height, $h = 1.8\,\text{m}$

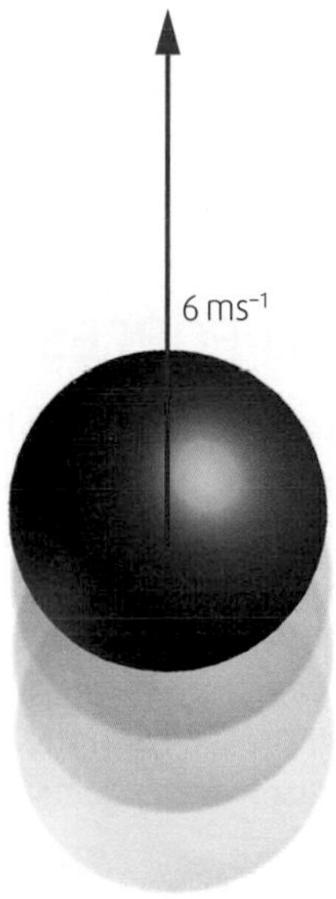

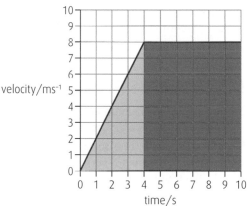
6 ms⁻¹

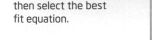
DON'T FORGET

List the *suvat* variables, then select the best fit equation.

DON'T FORGET

Remember to use + and – for vector quantities or the answer is likely to be wrong.

DON'T FORGET

An object is dropped downwards: If acceleration is negative, then so is displacement and the velocities.

CALCULATIONS FROM GRAPHS

5   Calculate the initial acceleration and total displacement of an object using data from the velocity-time graph shown.

$s = ?$
$u = 0\,\text{ms}^{-1}$
$v = 8\,\text{ms}^{-1}$
$a = ?$
$t = 4\,\text{s}$

$v = u + at$
$8 = 0 + (a \times 4)$
$a = 2\,\text{ms}^{-2}$
initial acceleration $= 2\,\text{ms}^{-2}$
$s = (0 \times 4) + (\frac{1}{2} \times 8 \times 4)$
$s = 16\,\text{m}$

velocity/ms⁻¹

time/s

DON'T FORGET

Use *s* for displacement, not *d*.

ONLINE

Follow the links at www.brightredbooks.net to explore this topic further.

THINGS TO DO AND THINK ABOUT

1   **Vectors**: *s*, *u*, *v* and *a* are vectors. *t* is not a vector.
2   **Directions**: Normally we will use these conventions:
    • upwards is positive(+), downwards is negative(−)
    • right or original direction is positive(+), left is negative(−).
3   **Gravity** pulls down whether an object is going up or down, so $a = g = -9.8\,\text{ms}^{-2}$ always.
4   In a **rocket**, acceleration up is positive due to thrust up.
5   Consider the statement "The ball accelerated into the goal". In a open question you can comment that the ball will only have accelerated towards the goal while in contact with the player's shoe but then gravity will have accelerated the ball towards the Earth while in flight. There are two independent values and directions.

VIDEO LINK

Check out the clip on equations of motion at www.brightredbooks.net

ONLINE TEST

Test yourself on calculations at www.brightredbooks.net

NEWTON'S 1ST AND 2ND LAWS

Newton's first two laws of motion apply to the relationship between balanced and unbalanced forces acting on an object and its motion.

BALANCED FORCES

Forces which are **equal in size** but **opposite in direction** are called **balanced forces**.

Stationary objects

A person sitting on a chair or a duck floating on water experience **balanced forces**. Both have **weight downwards**. The layers of atoms are compressed and the forces between the atoms exert a **normal force upwards** against the weight which results in a pair of balanced forces.

Objects in motion

A raindrop and a parachutist experience **balanced forces** when they fall to Earth with **constant (terminal) velocity**. The force which balances their **weight** is the force resulting from **air resistance** and there is no change in velocity. Note the raindrop has less weight and less air resistance whereas the parachutist has more weight and needs more air resistance. The large surface area of the parachute increases air resistance.

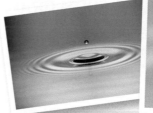

DON'T FORGET

More information on terminal velocity can be found on p 37.

NEWTON'S 1ST LAW OF MOTION

N1: An object will **remain at rest** or will **remain at constant velocity** unless acted on by an unbalanced force.

Newton's 1st law is a **no-force** law. In space travel, no force is needed for an object to keep moving. If there are no forces acting on an object, the object continues at a steady speed in a straight line.

Friction

Newton's 1st law is also a **no-unbalanced-force** law. The parachutist, the raindrop and the campervan all move at constant velocity when the forces are balanced out.

To keep a vehicle moving, we normally have to keep applying a force. Why is this?

An object in motion experiences **resistive forces** that **increase with velocity**. These are known as the forces of friction. **Friction** always acts **against** the **direction** of the **motion**.

Where the **applied force** and **friction** are **balanced**, then **Newton's 1st law** tells us that the object will remain at **constant velocity**.

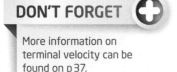

DON'T FORGET

Force is a vector.

contd

Aircraft

An aircraft is flying at a constant velocity. Draw a simple diagram of an aircraft and add labelled arrows to show the direction of the following four forces which are acting on it:

- thrust
- air resistance
- weight
- lift

As all the forces are balanced the aircraft maintains constant velocity.

DON'T FORGET

Balanced forces and no force gives the same effect.

DON'T FORGET

The force of **friction** increases with **velocity**.

UNBALANCED FORCES

The **resultant of forces** which do not cancel out is known as the **unbalanced force**.

An **unbalanced force** causes **acceleration**.

The dragster accelerates because the engine thrust is greater than the resistive forces.

Force is a vector quantity, so the resultant force equals the difference in the thrust and drag.

thrust   drag

NEWTON'S 2ND LAW OF MOTION

N2: The **acceleration** of an object **varies directly** with the **unbalanced force** and **inversely** with its **mass**.

$$a \propto \frac{F}{m} \qquad a = k\frac{F}{m}$$

The Newton defined

The unit of force is the **Newton**. **1 N** is defined as the resultant force, which causes a mass of **1 kg** to accelerate at **1 ms^{-2}**.

Substituting into the above equation: $1 = k\frac{1}{1}$

$$k = 1$$

$a = \dfrac{F}{m}$ or $F = ma$ known as Newton's 2nd law equation (**N2**).

An apple is thought to weigh about 1 N.

DON'T FORGET

The **unbalanced** force causes acceleration, so acceleration is fun : $F_{un} = ma$

Example:

A 100 kg vehicle accelerates. The engine force is 200 N, but friction exerts a force of 50 N in the opposite direction. Find the acceleration.

$$a = \frac{F_{un}}{m}$$

$$a = \frac{200 - 50}{100}$$

$$= 1.5 \, ms^{-2}.$$

Newton's 2nd law tells us that **force causes acceleration**.

ONLINE

Follow the links at www.brightredbooks.net to explore this topic further and take a mini quiz on this topic.

VIDEO LINK

Check out the clips at www.brightredbooks.net for more on Newton's 1st and 2nd laws.

ONLINE TEST

Head to www.brightredbooks.net and test yourself on Newton's 1st and 2nd laws.

THINGS TO DO AND THINK ABOUT

Rocket blast off!

Consider how the forces change as a rocket takes off. As the engines ignite, the weight will still be greater than the thrust, so there is no movement. The point of lift-off is the only point where Newton's 1st law applies (balanced forces). Only once some more fuel burns off will the rocket start to lift off and begin acceleration (unbalanced forces). As the fuel burns, mass and weight rapidly decrease, the acceleration keeps on increasing, until the engines are switched off. Under Newton's 2nd law, increasing unbalanced force cause increasing acceleration. Once the engines are off we might think the rocket will maintain a constant velocity but in reality a weak gravitational force will still cause a change in velocity. Gravitational force can also be used to maintain a satellite in orbit.

FREE BODY DIAGRAMS

Free body diagrams are used to help identify the forces involved in the motion of objects. The object(s) can be represented by a simple box with arrows added to identify the appropriate size and direction of the forces applied. Once this is done, resultant force and acceleration can be calculated.

ROCKETS

Adding the forces together gives the resultant force.

$$F = T + W$$

If **thrust value** is **positive** (up), **weight value** will be **negative** (down).

The weight value will vary.

Mass decreases as the fuel is rapidly used, so $W = mg$ decreases.

Gravitational field strength decreases with height so weight decreases.

The air resistance would be negative but can be considered as negligible.

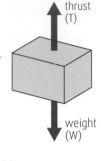

thrust (T)

weight (W)

Example:

A rocket of mass 9000 kg at take-off has a thrust of 180 000 N. Calculate the acceleration.

Solution:

$F = T + W = 180\,000 + (9000 \times -9\cdot8) = 91\,800\,N$

$a = \dfrac{F_{un}}{m} = \dfrac{91\,800}{9000} = 10\cdot2\,ms^{-2}$

upward force (U)

weight (W)

PERSON IN A LIFT

Upward force (U)   Let upward force be a positive number.

Downward force (W)   Let weight down be a negative number

Unbalanced force $F_{un} = U + W$

Acceleration $a = \dfrac{F_{un}}{m}$

If you stand on bathroom scales in a lift, your weight appears to change when the lift accelerates. Your weight doesn't change, but the scales read your weight plus the unbalanced force, causing acceleration (this gives the upthrust).

upward force (U)

weight (W)

Example:

1   If the lift is **stationary** or moving at a **constant speed**:
$F_{un} = U + W = 0\,N$
If mass = 70 kg, $W = mg = 70 \times -9\cdot8 = -686\,N$, so $U = +686\,N$
$F_{un} = U + W = 686 + (-686) = 0\,N \Rightarrow$ scales read 686 N

2   If the lift is **accelerating up**:
upward force increases
upward force > weight, so unbalanced force and acceleration are **positive**.
If the lift accelerates up at 2 ms⁻², accelerating force $F_a = ma = 70 \times 2 = 140\,N$
Total upward force = 686 + 140 = 826 N, scales read 826 N.

3   If the lift is **accelerating down**:
upward force decreases
upward force < weight, so unbalanced force and acceleration are **negative**.
If the lift accelerates down at 2 ms⁻², the scales will read 546 N.

TRAINS, CARAVANS AND PARCELS

Free body diagrams can be applied to **connected objects** that are being **pushed** or **pulled**. In these cases, it is necessary to include the tension between the connected objects.

Example:

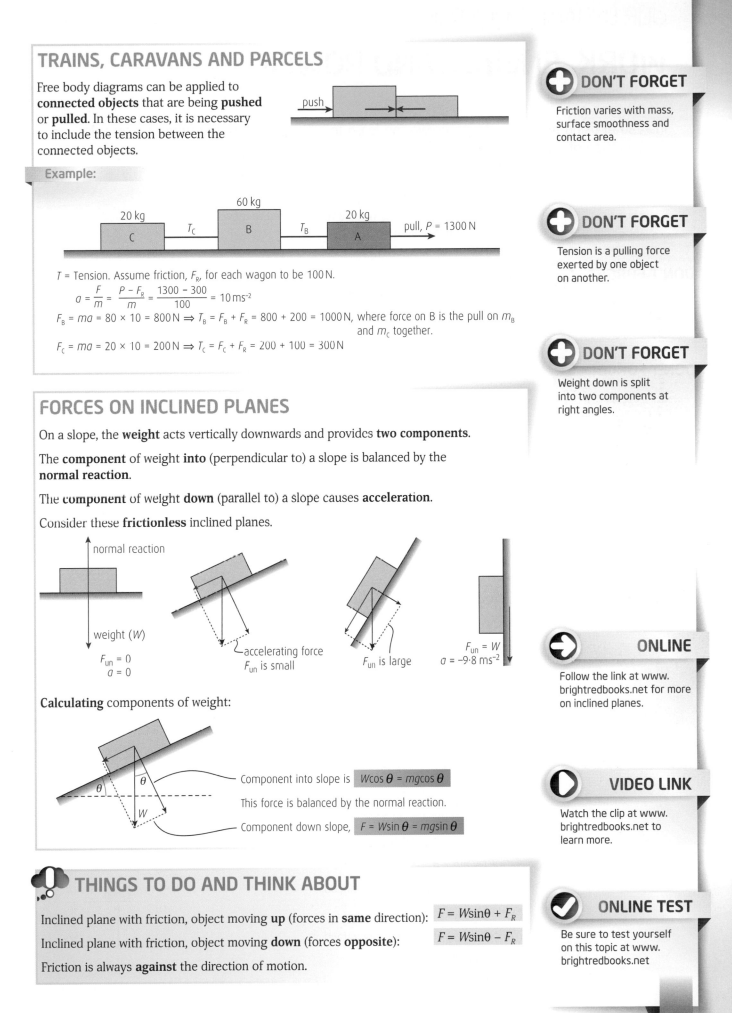

T = Tension. Assume friction, F_R, for each wagon to be 100 N.

$$a = \frac{F}{m} = \frac{P - F_R}{m} = \frac{1300 - 300}{100} = 10\,\text{ms}^{-2}$$

$F_B = ma = 80 \times 10 = 800\,\text{N} \Rightarrow T_B = F_B + F_R = 800 + 200 = 1000\,\text{N}$, where force on B is the pull on m_B and m_C together.

$F_C = ma = 20 \times 10 = 200\,\text{N} \Rightarrow T_C = F_C + F_R = 200 + 100 = 300\,\text{N}$

FORCES ON INCLINED PLANES

On a slope, the **weight** acts vertically downwards and provides **two components**.

The **component** of weight **into** (perpendicular to) a slope is balanced by the **normal reaction**.

The **component** of weight **down** (parallel to) a slope causes **acceleration**.

Consider these **frictionless** inclined planes.

Calculating components of weight:

Component into slope is $W\cos\theta = mg\cos\theta$

This force is balanced by the normal reaction.

Component down slope, $F = W\sin\theta = mg\sin\theta$

THINGS TO DO AND THINK ABOUT

Inclined plane with friction, object moving **up** (forces in **same** direction): $F = W\sin\theta + F_R$

Inclined plane with friction, object moving **down** (forces **opposite**): $F = W\sin\theta - F_R$

Friction is always **against** the direction of motion.

WORK, ENERGY AND POWER

Instead of using free body diagrams we can learn to solve problems on motion by considering conservation of energy and energy changes.

WORK, ENERGY AND POWER

Work is done on an object when a **force** is used to move the object a certain **distance**. Work is a **scalar** quantity.

$$E_W = Fd$$

Vertical component: $F_v = F\sin\theta$

Horizontal component: $F_h = F\cos\theta$

Work is only done when **energy is transferred**.

$$W = E_W$$

(Note that the symbols W and E_W may be interchanged in calculations.)

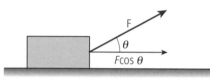

$$E_W = F\cos\theta \times d$$

Power is the **rate** of **doing work**.   Power is the **work done** in **unit time**.

$$P = \frac{E}{t}$$

$$P = \frac{E_W}{t} = \frac{Fd}{t} = Fv$$

$$P = Fv$$

Example:

A parcel is pulled along a distance of 5 m at a speed of 4 ms⁻¹ by a force of 50 N.

$E_W = Fd = 50 \times 5 = 250\,J$
$P = Fv = 50 \times 4 = 200\,W$

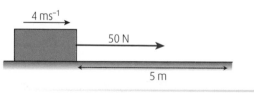

POTENTIAL ENERGY

Gravitational potential energy is the energy that an object has due to its position in a force field.

A mass is lifted against the gravitational field.

$$E_W = Fd = mg \times h$$

Lifting force and weight are balanced.

$$E_P = mgh$$

Work done = potential energy gained

Example:

A 5 kg mass is lifted through a height of 2 m. $E_p = mgh = 5 \times 9{\cdot}8 \times 2 = 98\,J$

KINETIC ENERGY

The **kinetic energy** of an object is the energy that it possesses due to its motion.

An **unbalanced force** applied over a **distance** causes **acceleration** and an increase in **kinetic energy**. Work is done and energy is transferred.

The gain in kinetic energy: $E_k = \frac{1}{2}mv^2$

An increase in speed of 0–60 mph takes four times as much energy as 0–30 mph. $E_k \propto v^2$

Example:

1 How much more energy is required to go to 90 mph compared with going to 30 mph?

2 How much energy is required for a 500 kg car to go to 20 ms⁻¹ from rest?

Solutions:

1 nine times

2 $E_k = \frac{1}{2}mv^2 = \frac{1}{2} \times 500 \times 20^2 = 100\,000\,J$

CONSERVATION OF ENERGY

Energy is never created or destroyed, just changed from one form into another.

When a car accelerates from rest to a high speed, **work** changes to **kinetic** energy.

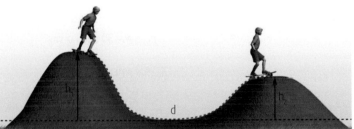

traffic lights

$Fd = \frac{1}{2}mv^2$ assuming no work is done against friction.

In real life, **friction** is **transferring energy to the surroundings**. Some work is done against friction, and the rest becomes kinetic energy.

$Fd = \frac{1}{2}mv^2 + F_{fr}d$

A ball is dropped from a height of 2 m. **Potential** energy is converted to **kinetic** energy.

$mgh = \frac{1}{2}mv^2$ assuming no air resistance.

A skateboarder skates from one high hill to a lower one, as shown.

The loss in **potential** energy is converted into **work done** against the frictional forces, and the remainder will be transferred to **kinetic** energy.

$E_p = E_W + E_k$

$mg(h_1 - h_2) = F_{fr}d + \frac{1}{2}mv^2$

If the skateboarder just reaches the top of hill 2, then E_k is zero.

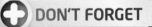 **DON'T FORGET**

In E_p and E_k equations, only the gain or loss in energy is calculated.

 **VIDEO LINK**

Head to www. brightredbooks.net for some great clips on work, energy and power.

THINGS TO DO AND THINK ABOUT

1 When energy is transferred to the **surroundings**, it is usually in the form of **heat**. Energy is said to be degraded, as it is hard to recover this energy.

2 The force of friction is the **cause** of energy lost to the surroundings.

 **ONLINE TEST**

Take the test on work, energy and power at www. brightredbooks.net

MOMENTUM, ENERGY AND EXPLOSIONS

Here we consider momentum and energy and learn what principles are revealed during explosions.

The Flying Scotsman

MOMENTUM

Momentum is a measure of the velocity and the mass of a moving object. Consider a person running along a platform alongside a train with identical velocity. Do they have the same or a different motion? The train has a greater mass, so the train has a greater momentum.

An object with more momentum requires a large amount of force to get it moving and a large amount of force is required to bring it to rest. A space ship moving quickly (having both large mass and large velocity) has a very large momentum.

We define momentum as the product of mass and velocity. The symbol for the momentum of an object is p. $p = mv$

The **units** of momentum are simply made from the product of the units of mass and velocity. $kg\,ms^{-1}$

We can use + and − to indicate direction, such as for right and left.

Consider a car of mass 1000 kg moving with a velocity whose magnitude is 30 ms⁻¹. Compare that with a lorry of mass 3000 kg moving in the opposite direction with a velocity of 10 ms⁻¹.

Lunar lander

Both have the same value of momentum but the opposite direction.

$p_{car} = +30\,000\,kg\,ms^{-1}$

$p_{lorry} = -30\,000\,kg\,ms^{-1}$

Example:

A tennis ball of mass 58 g is travelling with a velocity of 50 ms⁻¹. Calculate the value of its momentum.

Solution:

$p = mv = 0.058 \times 50 = 2.9\,kg\,ms^{-1}$

ENERGY

An object in motion has **kinetic energy**. This is the amount of energy needed to get the object to its travelling velocity and we assume it takes no more energy to keep it going. The same amount of energy is required to bring the object to rest.

Kinetic energy varies with the mass and the square of the velocity. $E_k = \frac{1}{2}mv^2$

Example:

How much kinetic energy does the tennis ball above have?

Solution:

$E_k = \frac{1}{2}mv^2 = \frac{1}{2} \times 0.058 \times (50)^2 = 72.5\,J$

EXPLOSIONS

An explosion provides a sudden release of energy. We will consider where this energy causes two objects to exert a force on each other and separate.

before   after

Explosion blows lorry apart

At the time of an explosion the objects may be stationary or may be travelling with some velocity. After the explosion, the velocities of the different objects depends on the masses of these objects as well as their initial velocities. Experiments show **momentum is conserved** even though the velocities are not.

The conservation of momentum is very useful in the **calculation** of the motion of objects after an explosion or where two objects exert a force on each other and separate. The ideas can be applied to a bullet leaving a rifle, an alpha particle leaving a radium nucleus, an arrow fired from a bow or a person leaving a boat to go ashore.

Conservation of momentum

During an explosion, the **law of conservation of linear momentum** can be applied to the **interaction** of **two objects** moving in **one dimension**, in the **absence** of net external **forces**.

Momentum is conserved, in the **absence** of net external **forces**.

Consider an explosion where an object breaks into two pieces:

Example:

Total momentum before = total momentum after
$mu = m_1v_1 + m_2v_2$
$(3 \times 0) = (2 \times -50) + (1 \times v_2)$
$0 = -100 + v_2$
$v_2 = 100\,\text{ms}^{-1}$ to the right (since v_2 is positive)

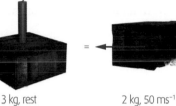

3 kg, rest   2 kg, 50 ms⁻¹   1 kg, v = ?

Dynamite fractures rock into two parts

Energy

During an explosion, we can also consider what has happened to energy. **Energy** has to be **put in** to cause the explosion. Chemical or potential energy changes to **kinetic energy**.

Consider the example above.

Kinetic energy before: E_k before $= \frac{1}{2}mu^2 = 0\,\text{J}$

Kinetic energy after: E_k after $= \frac{1}{2}m_1v_1^2 + \frac{1}{2}m_2v_2^2 = \frac{1}{2} \times 2 \times 50^2 + \frac{1}{2} \times 1 \times 100^2 = 7500\,\text{J}$

The **kinetic energy** after is **greater** than before, so E_k **is not conserved**.

The difference in the **kinetic energy** before and after is equivalent to the amount of **potential energy** supplied **by the explosive** during the explosion.

For an **explosion**, we should always calculate that there is more kinetic energy **after** than before.

💭 THINGS TO DO AND THINK ABOUT

Estimate the mass of a rifle and a bullet. As you learn to be a physicist you should learn to make reasonable estimates of data. Assume the rifle recoil speed is 2 ms⁻¹. Calculate the speed of the bullet. Ask yourself if your answer seems reasonable. At Higher level you should always check to see if your answer appears reasonable. Sometimes when we work on the maths we get impossible answers by a simple mistake.

DON'T FORGET

In the lab, a **compressed spring** can release its potential energy and provide trolleys with a sudden increase in kinetic energy.

ONLINE

Explore this further by following the link at www.brightredbooks.net

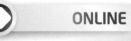

VIDEO LINK

Watch the clip at www.brightredbooks.net for more about momentum in explosions.

DON'T FORGET

In an **explosion**, momentum is conserved but kinetic energy is not as it increases.

ONLINE TEST

Head to www.brightredbooks.net and take the test on momentum, energy and explosions.

MOMENTUM, ENERGY AND COLLISIONS

Here we consider momentum and energy and learn what principles are revealed during collisions.

 +  =

INELASTIC COLLISIONS

During an inelastic collision, objects make contact and in a perfect inelastic collision, stick together to form a single object. An inelastic collision can be considered as an explosion in reverse.

Consider two lorries. The lorry moving to the right is moving at $20\,ms^{-1}$ and has a total mass of 8000 kg. The lorry moving to the left is empty, is moving at $30\,ms^{-1}$ and has a mass of 5000 kg. Assuming that the lorries collide head on and stick together, we can calculate their combined velocity after the collision.

DON'T FORGET

Momentum is a vector. The velocities are positive or negative.

Momentum

In an **inelastic** collision, **momentum is conserved**, in the absence of net external forces.

We can use this law of conservation to calculate the combined velocity after the collision above. Note that the lorries are assumed to have the same velocity after this collision. There are two similar methods.

Method 1

Consider the objects as one after the collision.

Total momentum before = total momentum after

$m_1 u_1 + m_2 u_2 = m_{12} v_{12}$

$(8000 \times 20) + (5000 \times -30) = (13\,000 \times v_{12})$

$160\,000 - 150\,000 = 13\,000 \times v_{12}$

$v_{12} = 0.77\,ms^{-1}$

The lorries will move together with a combined velocity of $0.77\,ms^{-1}$ to the right.

Method 2

Consider the objects as two parts with the same velocity.

Total momentum before = total momentum after

$m_1 u_1 + m_2 u_2 = m_1 v_{12} + m_2 v_{12}$

$(8000 \times 20) + (5000 \times -30) = (8000 \times v_{12}) + (5000 \times v_{12})$

$160\,000 - 150\,000 = 13\,000 \times v_{12}$

$v_{12} = 0.77\,ms^{-1}$

The lorries will move together with a combined velocity of $0.77\,ms^{-1}$ to the right.

You can also work out the momentum of each part before and after individually using $p = mv$, then use conservation of momentum to complete the question. You must remember that momentum is a vector and use appropriate + and – signs.

Energy in inelastic collisions

In an **inelastic** collision, **kinetic energy** is **not conserved**. Energy is **given out**.

Some kinetic energy changes to heat and sound energy. Consider the trucks example above.

contd

Kinetic energy before

E_k before $= \frac{1}{2} m_1 u_1^2 + \frac{1}{2} m_2 u_2^2$

$= \frac{1}{2} \times 8000 \times (20)^2 + 5000 \times (-30)^2$

$= 1\,600\,000 + 450\,000$

$= 2\,050\,000\,J$

Kinetic energy after

E_k after $= \frac{1}{2} m_{12} v_{12}^2$

$= \frac{1}{2} \times 13\,000 \times (0\cdot77)^2$

$= 3854\,J$

DON'T FORGET

Energy is a scalar quantity.

The **kinetic energy** after is **less than** before. **Kinetic energy** is shown to be **not conserved**.

Due to the friction on impact $2\,046\,146\,J$ of energy have been lost from the trucks to heat and sound.

ELASTIC COLLISIONS

An **elastic collision** is one where both **momentum** and **kinetic energy** are **conserved**.

Elastic collisions take place between magnets and charged particles.

Momentum

Total momentum before = total momentum after $m_1 u_1 + m_2 u_2 = m_1 v_1 + m_2 v_2$

Energy

Energy is **conserved**.

DON'T FORGET

A question may require you to use either the equation for momentum or for energy. They are not the same quantities, and you may need to calculate both.

Kinetic energy before $E_{k\,before} = \frac{1}{2} m_1 u_1^2 + \frac{1}{2} m_2 u_2^2$

Kinetic energy after $E_{k\,after} = \frac{1}{2} m_1 v_1^2 + \frac{1}{2} m_2 v_2^2$

ONLINE

Investigate the link at www. brightredbooks.net to learn how to determine whether a collision is elastic or not.

If the total kinetic energies **before** and **after** are the same, then we have an **elastic collision**.

$E_{k\,before} = E_{k\,after} \Rightarrow \frac{1}{2} m_1 u_1^2 + \frac{1}{2} m_2 u_2^2 = \frac{1}{2} m_1 v_1^2 + \frac{1}{2} m_2 v_2^2$

Example:

A linear air-track vehicle, mass = 400 g and moving through a light gate at $2\cdot0\,ms^{-1}$, collides with a stationary vehicle, mass = 600 g, which a second light gate records as travelling off at $1\cdot6\,ms^{-1}$.

Both vehicles were fitted with magnets, and they did not appear to touch during the collision. Calculate the velocity of the first vehicle through one of the light gates after the collision. Show what type of collision this was.

400 g $2\cdot0\,ms^{-1}$ 600 g $1\cdot6\,ms^{-1}$

Solution:

Total momentum before = total momentum after

$m_1 u_1 + m_2 u_2 = m_1 v_1 + m_2 v_2$

$(0\cdot4 \times 2\cdot0) + (0\cdot6 \times 0) = (0\cdot4 \times v_1) + (0\cdot6 \times 1\cdot6)$

$v_1 = -0\cdot4\,ms^{-1}$

Velocity of vehicle 1 after is **$0\cdot4\,ms^{-1}$** to the **left**.

Kinetic energy before $E_{k\,before} = \frac{1}{2} m_1 u_1^2 + \frac{1}{2} m_2 u_2^2$ $= \frac{1}{2} \times 0\cdot4 \times 2\cdot0^2 + 0 = 0\cdot8\,J$

Kinetic energy after $E_{k\,after} = \frac{1}{2} m_1 v_1^2 + \frac{1}{2} m_2 v_2^2$ $= \frac{1}{2} \times 0\cdot4 \times 0\cdot4^2 + \frac{1}{2} \times 0\cdot6 \times 1\cdot6^2 = 0\cdot8\,J$

No change in kinetic energy $\Rightarrow$ **elastic collision**

VIDEO LINK

Check out the clip at www. brightredbooks.net for more on collisions and momentum conservation.

THINGS TO DO AND THINK ABOUT

Momentum is conserved.
Momentum is a vector.
Kinetic energy is only conserved in elastic collisions. **Kinetic energy** is given out from inelastic collisions and taken in during explosions.
Total energy is always conserved.

ONLINE TEST

Head to www. brightredbooks.net and take the test on momentum, energy and collisions.

NEWTON'S 3RD LAW

Newton's 3rd law tells us that forces exist in **pairs**. These **Newton pairs** are **equal in size** but **opposite in direction**. Newton's 3rd law was originally written as "For every action there is an equal and opposite reaction". This can suggest some time delay between the two forces, but it is important to realise that these two forces occur in pairs **at the same time**.

N3: If A exerts a force on B, then B exerts an **equal but opposite** force on A.

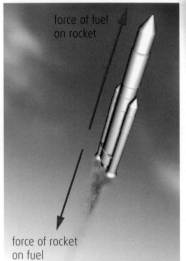

force of fuel on rocket

force of rocket on fuel

APPLYING THE 3RD LAW OF MOTION

In a rocket, the **rocket** pushes the **fuel** away, while the **fuel** pushes the **rocket** away.

Both forces occur at the **same time**.

Pairs of forces are all around us. Forces do not exist singly, because there is always an opposite force. If A exerts a force on B, then the reverse **must** happen.

These ice skaters push away together. What would be the result if their mass were different? The larger mass would accelerate less even though the same size of force has been applied.

What happens when one person tries to push his friend away? They both accelerate away. What happens if his friend grabs hold and tries to remain? They both accelerate together towards each other.

Note we have used Newton's 2nd law to explain the effect of each individual force on each object (each object accelerates), even though there is a pair of forces acting.

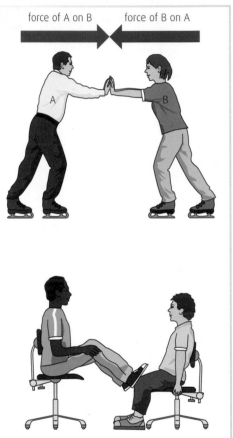

force of A on B   force of B on A

DON'T FORGET

An object will travel at constant velocity when all the forces on it are balanced.

FRICTION

A **car** exerts a force on the **road** so, according to Newton's **3rd** law, the **road** exerts an equal and opposite force on the **car**. This opposite force, according to Newton's **2nd** law, makes the car **accelerate**. Why does the car not continue to accelerate but reach a **constant velocity**? For a constant velocity, according to Newton's **1st** law, there must be **balanced** forces. This is where we need to consider the frictional forces acting against the motion of the car.

Frictional force, a negative vector quantity which always opposes motion, increases with

- velocity
- surface area
- contact pressure between the surfaces.

Friction force in the diagram will include air resistance, friction in the wheel bearings and resistance between the tyres and the road.

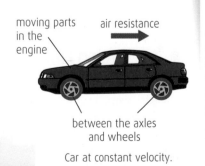

moving parts in the engine

air resistance

between the axles and wheels

Car at constant velocity.

TERMINAL VELOCITY

A skydiver descends

Consider the forces acting on a skydiver during the descent.

First, we consider that the only force acting on the skydiver is his weight or the force of gravity. Newton's 2nd law tells us this will cause acceleration and we know its initial value will be $9.8\,\mathrm{ms^{-2}}$. The weight, we assume, will remain constant throughout the dive although if you were about to make a record attempt from extremely high in the atmosphere it will theoretically start slightly less. Weight, $W = mg$ and acts downwards.

As the skydiver descends, the air resistance will increase and the resultant downward force decreases, so the acceleration decreases although the velocity is still increasing.

We can see this in the first section of the velocity-time graph. You can estimate the acceleration by considering the change in velocity over any one-second time interval.

Try to estimate the acceleration over the 1st second, the 10th second and the 20th second. You will see that the values of acceleration are decreasing.

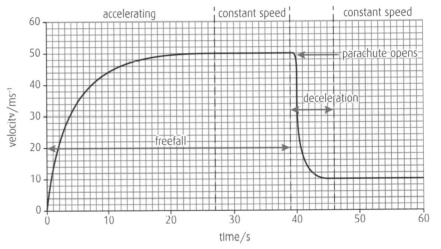

The constant speed reached when the air resistance balances the weight is known as the **terminal velocity**. It does not change until the parachute is opened to increase the surface area and the air resistance. When the parachute is opened, the velocity decreases (though the skydiver is still going downwards) so the air resistance decreases again till the forces are again balanced and a new lower terminal velocity is reached.

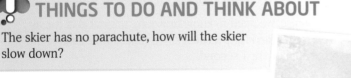

THINGS TO DO AND THINK ABOUT

The skier has no parachute, how will the skier slow down?

ONLINE

Follow the link at www. brightredbooks.net to learn more about Newton's 3rd law and aerodynamics.

VIDEO LINK

Check out the clips 'Skydiving without a parachute!', 'Space jump world record 2012' and 'Newton's cradle' at www. brightredbooks.net

ONLINE TEST

Test yourself on Newton's 3rd law at www. brightredbooks.net

N3 AND IMPULSE

NEWTON'S 3RD LAW

The table summarises the applications of the law of conservation of momentum for explosions and collisions.

TO AN EXPLOSION FROM REST		FOR A COLLISION
total momentum before = total momentum after		total momentum before = total momentum after
$0 = m_1v_1 + m_2v_2$	(see **a** below)	$m_1u_1 + m_2u_2 = m_1v_1 + m_2v_2$
$m_1v_1 = -m_2v_2$		$m_1(v_1 - u_1) = -m_2(v_2 + u_2)$
$\dfrac{m_1v_1}{t} = \dfrac{-m_2v_2}{t}$	same contact time	$\left[\dfrac{m_1(v_1 - u_1)}{t} = \dfrac{-m_2(v_2 - u_2)}{t}\right]$
$m_1a_1 = -m_2a_2$		$m_1a_1 = -m_2a_2$
$F_1 = -F_2$	Newton's 3rd law (see **b** below)	$F_1 = -F_2$

The law of conservation of momentum shows us that:

a the changes in momentum of each object are equal in size and opposite in direction
b the forces acting on each object are equal in size and opposite in direction.

IMPULSE

An object is accelerated by a force F for a time t.

$$F = ma = \frac{m(v - u)}{t} = \frac{mv - mu}{t}$$

Unbalanced force = rate of change of momentum

This is how Newton first described his Second Law of motion.

Force gives the rate of change, but not the change (of momentum). Rearrange the equation further:

$$Ft = mv - mu$$

- The product Ft is called the **impulse** and is the cause of the change in motion. Impulse is measured in Ns or kgms^{-1}.
- The change in momentum $mv - mu$ is the effect of the impulse.

To understand the concept of impulse, we need to know how long a force acts for, as well as the size of the force. A force applied for 5 s causes five times more change in momentum than the same force applied for 1 second.

A change in momentum depends on

- the size of force, and
- the time the force acts.

Example:

A force of 100 N is applied to a small 150 g ball for 0·020 s. Find the final velocity.

Solution:

$Ft = mv - mu$
$100 \times 0·020 = 0·150 \times (v - 0)$ velocity = 13·3 ms^{-1} in the direction of force

IMPULSE FROM A GRAPH

A small force applied for a long time causes the same change in momentum as a large force applied for a short time. This effect is used for crumple zones in cars.

Impulse = area under a $\frac{F}{t}$ graph

In real situations, the force is not usually constant. We must either

- calculate the average force × time, $\bar{F}t$, or
- calculate the area under a $\frac{F}{t}$ graph.

longer collision ⇒ less force

average force   (Impulse = area)

Change of momentum

Consider a ball deforming as it makes contact with a wall.

Change in momentum = $p_{final} - p_{initial} = (-mv) - (mv) = -2mv$

i.e. a change of $2mv$ towards the left.

hard ball

soft ball

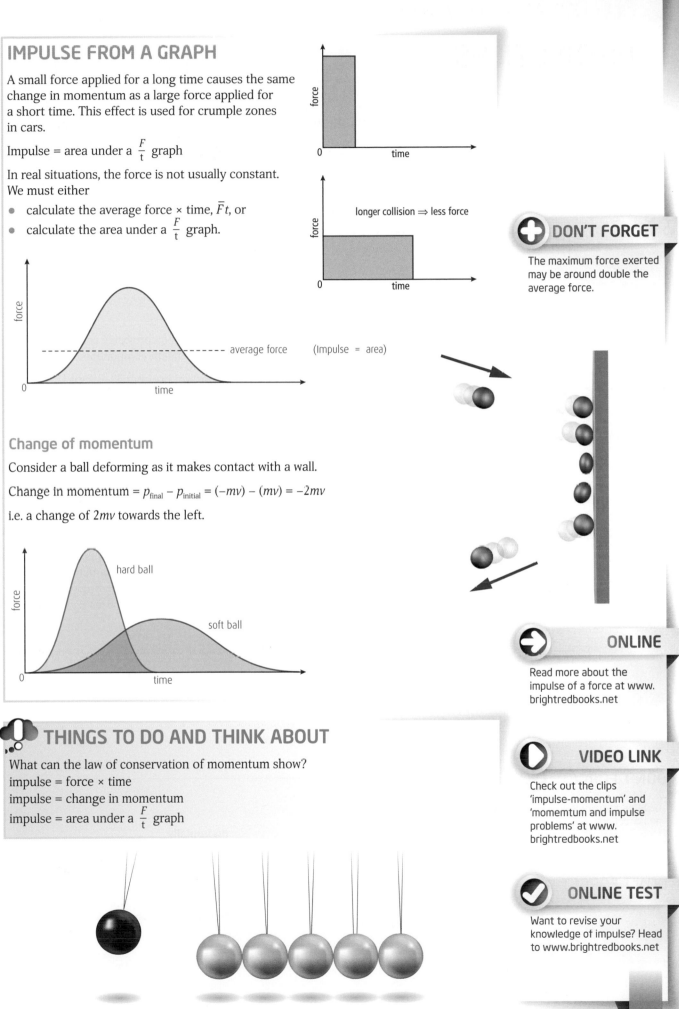

THINGS TO DO AND THINK ABOUT

What can the law of conservation of momentum show?
impulse = force × time
impulse = change in momentum
impulse = area under a $\frac{F}{t}$ graph

PROJECTILES

Previous problems have investigated objects thrown up, objects dropped vertically and objects going straight on a slope. Another type of motion results when an object is projected horizontally or at an angle. This type of motion is known as **projectile motion**.

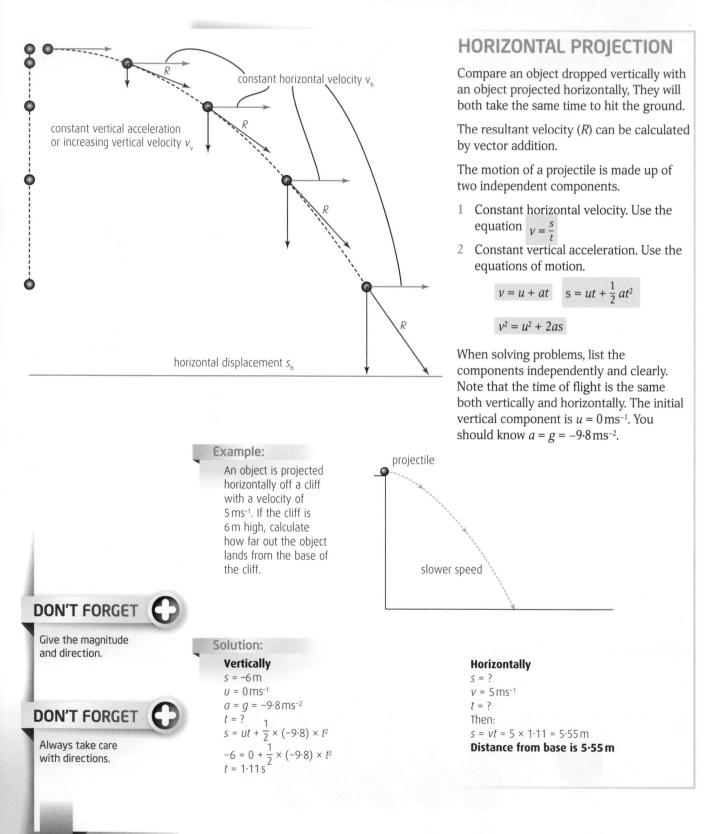

constant horizontal velocity v_h

constant vertical acceleration or increasing vertical velocity v_v

horizontal displacement s_h

HORIZONTAL PROJECTION

Compare an object dropped vertically with an object projected horizontally, They will both take the same time to hit the ground.

The resultant velocity (R) can be calculated by vector addition.

The motion of a projectile is made up of two independent components.

1   Constant horizontal velocity. Use the equation $v = \dfrac{s}{t}$

2   Constant vertical acceleration. Use the equations of motion.

$$v = u + at \qquad s = ut + \frac{1}{2}at^2$$

$$v^2 = u^2 + 2as$$

When solving problems, list the components independently and clearly. Note that the time of flight is the same both vertically and horizontally. The initial vertical component is $u = 0\,\text{ms}^{-1}$. You should know $a = g = -9\cdot8\,\text{ms}^{-2}$.

Example:

An object is projected horizontally off a cliff with a velocity of $5\,\text{ms}^{-1}$. If the cliff is 6 m high, calculate how far out the object lands from the base of the cliff.

projectile

slower speed

DON'T FORGET

Give the magnitude and direction.

DON'T FORGET

Always take care with directions.

Solution:

Vertically
$s = -6\,\text{m}$
$u = 0\,\text{ms}^{-1}$
$a = g = -9\cdot8\,\text{ms}^{-2}$
$t = ?$
$s = ut + \dfrac{1}{2} \times (-9\cdot8) \times t^2$
$-6 = 0 + \dfrac{1}{2} \times (-9\cdot8) \times t^2$
$t = 1\cdot11\,\text{s}$

Horizontally
$s = ?$
$v = 5\,\text{ms}^{-1}$
$t = ?$
Then:
$s = vt = 5 \times 1\cdot11 = 5\cdot55\,\text{m}$
Distance from base is 5·55 m

PROJECTION AT AN ANGLE

An object may be projected into the air at an angle. For example, an arrow is fired at an angle from the horizontal.

The initial velocity of the projectile must be resolved into horizontal and vertical components.

$$u_h = u\cos\theta \quad u_v = u\sin\theta$$

The horizontal component of velocity is constant.

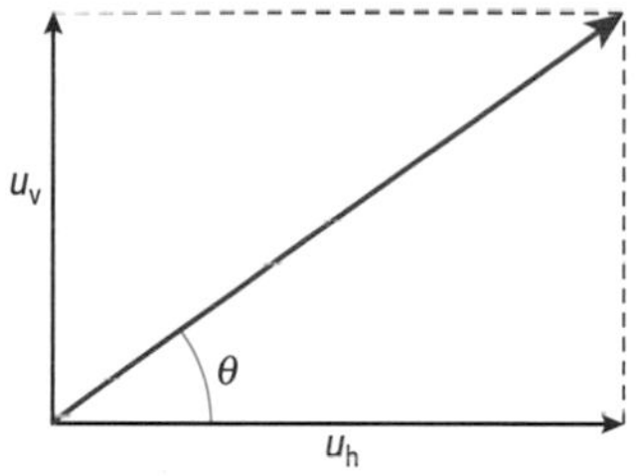

There is constant vertical acceleration due to the force of gravity acting downwards. Assuming air resistance is negligible, the trajectory will be symmetrical.

- The final vertical component of velocity = – the initial vertical component of velocity.
  $v_v = -u_v$
- The time of flight = 2 × time to the highest point.
- The vertical component of velocity at the top = $0\,\text{ms}^{-1}$ (but still has horizontal component).
- Vertically, $a = g = -9{\cdot}8\,\text{ms}^{-2}$.

Example:

A golfer strikes a ball which moves off at an angle of 25° to the ground at $60\,\text{ms}^{-1}$. The ball lands 6 s later. What distance does the ball travel and what is the vertical component of its velocity just before it hits the ground?

Solution:

Horizontally: $u_h = u\cos\theta = 60 \times \cos25° = 54{\cdot}4\,\text{ms}^{-1}$
$\qquad\qquad s = vt = 54{\cdot}4 \times 6 = 326{\cdot}4\,\text{m} \qquad$ distance = $326{\cdot}4\,\text{m}$
Vertically: $u_v = u\sin\theta = 60 \times \sin25° = 25{\cdot}4\,\text{ms}^{-1}$ upwards.
Final vertical component of velocity = $25{\cdot}4\,\text{ms}^{-1}$ downwards

THINGS TO DO AND THINK ABOUT

The **horizontal** and the **vertical components** are completely **independent**. That is, one does not affect the other. Nothing, except air resistance, changes the horizontal component, but gravity changes vertical motion. The time is, however, the same for both components.

ONLINE

Head to www.brightredbooks.net for more information on projectiles.

DON'T FORGET

The cosine component is adjacent to the angle.

DON'T FORGET

Remember to keep horizontal and vertical components separate.

ONLINE

Read an introduction to projectiles at www.brightredbooks.net

VIDEO LINK

Watch the video clips about projectiles at an angle at www.brightredbooks.net

ONLINE TEST

Head to www.brightredbooks.net and sit the test on projectiles to check your knowledge.

GRAVITATION

Ideas about gravity and gravitational fields have been fundamental to physics for more than 300 years since Isaac Newton developed some great thinking in this area.

FIELDS

DON'T FORGET

A mass will move in the **direction** indicated by the field.

Fields are a concept used to explain the fact that a **force** can be exerted on an object at a distance. A **mass** released at a distance from the Earth will experience a force in the direction towards the Earth. We say the object is in a gravitational field and use lines with arrows to illustrate both the **strength** and **direction** of the field.

In a **gravitational field**, a **mass** experiences a **force**.

The **gravitational field strength** is a measure of the **force** on a **unit mass**.

Uniform fields

In most experiments and experiences **on Earth** we consider the **gravitational field** to be **constant in both strength and direction**. From the deepest glen to the highest mountain the gravitational field strength is taken as

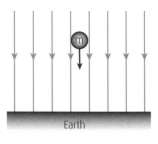

Earth

$$g = 9.8 \pm 0.05\,\text{Nkg}^{-1}$$

Gravity's direction does not vary by 1° vertically until we have travelled about 70 miles around the Earth's surface and so can be taken as constant over a short distance and **vertically downwards**.

Radial fields

DON'T FORGET

Any spherical object with uniform density behaves like a point source with a radial field outside the object.

When we consider space physics our scale of thinking changes. Objects can be at vast distances from each other and even large masses such as the planets can be considered as **point sources**. The **gravitational field decreases** as we move away from the Earth and the direction is **radial** like the spokes on a bicycle.

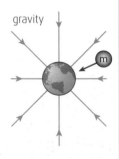

gravity

GRAVITY

Gravity or **Gravitational Field Strength (g)** is defined as the **force** exerted on **unit mass**.

$$g = \frac{F}{m}$$

DON'T FORGET

Field lines arrive at the surface at 90° to the surface.

We can consider unit mass as a test mass equal to 1 kg. Thus the force exerted on 1 kg of mass is a measure of the gravitational field strength. On Earth this force is 9·8 N. This is why **g = 9·8 Nkg⁻¹**.

Example:

Our bodies have more mass than 1 kg so our weight is found by multiplying g by the number of kilograms of mass in our body. This is in accordance with $W = mg$. If my mass is 82 kg, then my weight is

$W = mg = 82 \times 9.8 = 803.6\,\text{N}$. Note when g is fixed my weight varies with my mass.

Equivalence of acceleration with g

DON'T FORGET

The equation $s = ut + \frac{1}{2}at^2$ can be written as $s = ut + \frac{1}{2}gt^2$ in order to calculate g in a motion experiment.

Weight is a force, so we can show the equivalence of a with g.

$$a = \frac{F}{m} = \frac{W}{m} = \frac{mg}{m} = g$$

Earth has a gravitational field strength of $g = 9.8\,\text{Nkg}^{-1}$ so every object will travel downwards with acceleration, $a = 9.8\,\text{ms}^{-2}$.

NEWTON'S THOUGHT EXPERIMENT

Isaac Newton studied the effect of gravity on projectiles. He arrived at his famous **thought experiment**. Imagine a cannonball fired horizontally. The ball has a **constant horizontal velocity** but also a **constant vertical acceleration** which brings it down to the ground. Now Newton imagined what would happen if the ball could be fired with a greater and **greater horizontal speed**. The ball will travel further each time but the curvature of the Earth will start to be noticed. Although the ball is always accelerating towards the ground, due to gravity, a point will come when the ball will never reach the ground but be in orbit around the Earth.

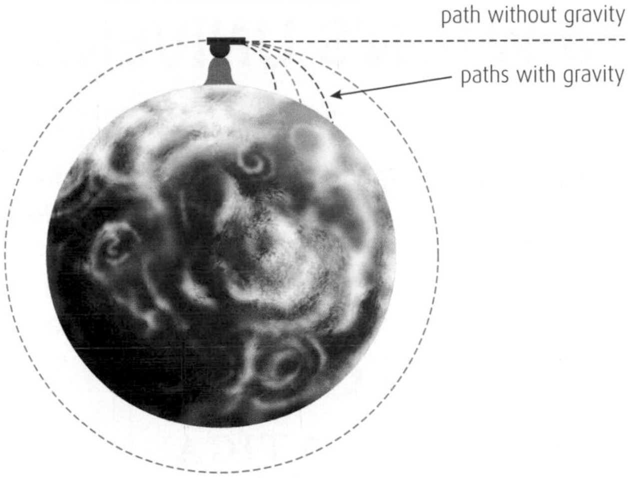

If there was no gravity the ball would travel in a straight line out into space.

The path without gravity is obeying Newton's 1st law of motion.

The path with gravity is obeying Newton's 2nd law of motion.

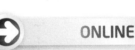
path without gravity
paths with gravity

Newton's thought experiment

Satellites

The cannonball in Newton's day could not be launched high enough or fast enough to go into orbit. If it had been it would have been the world's first artificial satellite. Satellites remain in orbit because the gravitational field keeps them accelerating towards the Earth. The Earth's surface just falls away at the same rate.

Astronauts in the International Space Station are not weightless. They experience gravity in the same way the station does and both are in the same orbit giving an apparent weightless situation.

THINGS TO DO AND THINK ABOUT

Since 1957, when the Russians launched Sputnik 1, more than 8000 artificial satellites have been launched. Find out more about why and how these have been launched.

DON'T FORGET

Gravity makes the apple fall to the ground and keeps satellites in orbit.

ONLINE

Find out more about radial fields and g at www.brightredbooks.net

VIDEO LINK

Check out the video about Newton's thought experiment at www.brightredbooks.net

VIDEO LINK

Watch the clip about geostationary satellites at www.brightredbooks.net

ONLINE TEST

Take the test at www.brightredbooks.net to check how well you have learned about gravitation.

NEWTON'S UNIVERSAL GRAVITY

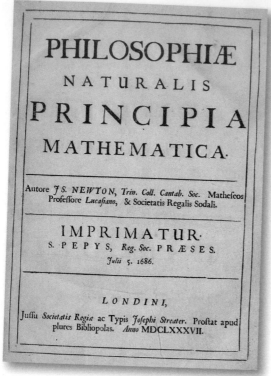

In 1687, Isaac Newton published his *Philosophiæ Naturalis Principia Mathematica*, (usually referred to simply as *Principia*). This is one of the most important books in science. In it, he gave his three laws of motion and his law of universal gravitation which we will study in depth here.

INVERSE SQUARE LAW OF GRAVITATION

Gravity is familiar to us as a force on mass that attracts objects such as an apple to the ground, giving them an **acceleration**. Newton not only realised this but he also realised that it was this **same force** that was keeping the Moon in **orbit** around the Earth. In *Principia* he went further.

Newton observed or demonstrated that every object (or mass) in the whole universe attracts every other object and this is responsible for the motion of the planets. We can see that gravity on Earth is just a particular case of **Newton's law of universal gravitation**.

Consider two objects (such as the Earth and the Moon) which are a large distance apart.

Newton combined ideas on $F \propto m$ and his third law to show that the force between two objects depended on both masses:

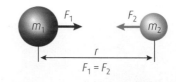

$F \propto m_1$

$F \propto m_2$

So $F \propto m_1 m_2$

Newton was also able to demonstrate (using Earth - Moon data) that the force between these two masses would also vary with the **inverse of the square of the distance** between the centres of the masses: $F \propto \dfrac{1}{r^2}$

These ideas are combined to give $F \propto \dfrac{m_1 m_2}{r^2}$

The universal gravitational constant

A constant is required to complete Newton's law but he did not have the data to allow him to calculate the value in his lifetime. With mass in kilograms, distance in metres and force in Newtons, the universal gravitational constant is calculated to be

$G = 6 \cdot 67 \times 10^{-11} \, \text{Nm}^2\text{kg}^{-2}$

The universal gravitational constant was first experimentally calculated in 1798 by Henry Cavendish. The value was improved by C.V. Boys in 1895. The gravitational force is very small and Cavendish and Boys both used very sensitive apparatus based on a torsion balance.

Newton's law of universal gravitation

The complete equation for gravitational force is $F = G \dfrac{m_1 m_2}{r^2}$

This law applies to

- point masses
- spheres of uniform density

and where other masses are far enough away for their effect to be negligible.

DON'T FORGET

Gravitational force is the weakest type of force.

NATURAL SATELLITES

The Moon is in orbit around the Earth and the planets are in orbit around the Sun. These orbits are maintained by the gravitational forces.

Example:

Calculate the gravitational force in an Earth-Moon system.

Solution:

$G = 6.67 \times 10^{-11}\,Nm^2kg^{-2}$
mass of Earth, $m_E = 6.0 \times 10^{24}\,kg$
mass of Moon, $m_M = 7.3 \times 10^{22}\,kg$
Earth to Moon distance, the radius of orbit, $r = 3.84 \times 10^8\,m$

$$F = G\frac{m_E m_M}{r^2} = 6.67 \times 10^{-11}\frac{(6.0 \times 10^{24})(7.3 \times 10^{22})}{(3.84 \times 10^8)^2} = 1.98 \times 10^{20}\,N = 2.0 \times 10^{20}\,N\ (2\ sig.\ figs.)$$

The Earth exerts this force on the Moon to keep it in orbit but the Moon also, in accordance with Newton's 3rd law, also exerts the same size of force on the Earth. The Earth is approximately 10^2 times more massive than the Moon but we can see the effect in the tides.

Example:

Calculate the force between the Earth and a man standing at the equator. Here we assume the Earth's mass is concentrated at its centre and the man in orbit is at the Earth's radius.

Solution:

$G = 6.67 \times 10^{-11}\,Nm^2kg^{-2}$
mass of Earth, $m_E = 6.0 \times 10^{24}\,kg$
mass of man, $m_m = 80\,kg$ (estimated)
Earth to man distance, the radius at the equator, $r = 6.4 \times 10^6\,m$

$$F = G\frac{m_E m_M}{r^2} = 6.67 \times 10^{-11}\frac{(6.0 \times 10^{24})(80)}{(6.4 \times 10^6)^2} = 781.6\,N = 780\,N\ (2\ sig.\ figs.)$$

Compare this with $F = mg = 80 \times 9.8 = 784\,N = 780\,N$ (2 sig. figs.)

The Earth exerts a force on the man and the man exerts this force on the Earth. You will know why the man accelerates more than the Earth.

Example:

Planets have large masses. Compare with two spheres of mass 1 kg each sitting on a smooth table 1 m apart. What force do they exert on each other?

Solution:

$$F = G\frac{m_1 m_2}{r^2} = 6.67 \times 10^{-11}\frac{(1)(1)}{(1)^2} = 6.7 \times 10^{-11}\,N\ (2\ sig.\ figs.)$$

We can see that the force is so small there is no way the two spheres will overcome friction and roll towards each other.

THINGS TO DO AND THINK ABOUT

Although Newton created these laws he struggled with the idea of a force being exerted on another object at a distance. Nowadays we can use Einstein's theory of general relativity where curved space-time is used to explain this effect created by mass.

DON'T FORGET

Objects such as charged particles and atoms have non-significant gravitational forces between them as the electrical and nuclear forces are much stronger.

DON'T FORGET

G is a universal constant, but g varies with planets and distance from the planet.

DON'T FORGET

The law of universal gravitation can be used from the smallest particles to the largest objects

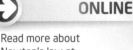

ONLINE

Read more about Newton's law at www.brightredbooks.net

VIDEO LINK

Watch the video clips at at www.brightredbooks.net for more on Newton's law and Newton to Einstein.

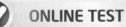

ONLINE TEST

Test yourself on Newton's universal gravity at www.brightredbooks.net

FORCES, PLANETS AND THE UNIVERSE

Here we will learn essential facts about universal gravitation and gravity.

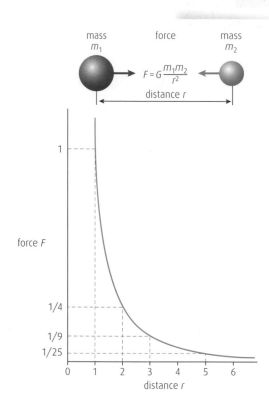

mass m_1 force   mass m_2

$F = G\dfrac{m_1 m_2}{r^2}$

distance r

force F

distance r

FORCES

The **force of gravity** acts on **mass**. The relative strength is very weak compared with the **electromagnetic force** which acts on **charges** or **moving charges**. You will find it is about 10^{35} times smaller. And the **strong nuclear force** is about 1000 times greater than the electromagnetic force. The strong nuclear force exists **inside the nucleus** and so does not come into our everyday consideration but it is considered in **particle physics**.

The Earth, other planets, the stars and galaxies have very **large masses** and are considered **electrically neutral**. This allows the gravitational field theory to be considered throughout the **universe**. A key development by Newton was the link between the effect of Earth's gravity on everyday objects and the **same gravity** being responsible for acting on the Moon and the planets.

However although Newton realised that gravity reached far out into space he also realised that its strength dropped dramatically with distance from an object. He even **invented calculus** to help in his calculations.

Newton demonstrated that gravitational field strength varied with the **inverse of the distance squared**. We can see from the graph or by calculation that if our gravitational field strength of $9{\cdot}8\,\text{Nkg}^{-1}$ gives 1 unit of force at the Earth's surface, then at twice that distance from the centre of the Earth the force is $\frac{1}{4}$, and at three times that distance from the centre of the Earth it has fallen to $\frac{1}{9}$ its value.

What is its relative value at six times the original distance?

$6^2 = 36$ so the relative value is $\frac{1}{36}$th.

g on planets

The inverse square graph always has the same shape, and can be used for the gravitational field round any isolated mass. The force at the surface on every kilogram of matter can be found from a table of gravitational field strengths for the planets. Our kilogram of matter is a test mass replacing m_2 in the diagram. We can see how the force on an object drops quickly with distance. However objects such as the International Space Station at about 400 km above the Earth's surface are relatively close to the Earth and the gravitational field value at that distance is still close to the value of Earth, $g = 8{\cdot}7\,\text{N kg}^{-1}$.

Our Sun and planetary data

Planet/Sun	Earth	Jupiter	Mars	Mercury	Moon	Neptune	Saturn	Sun	Uranus	Venus
Gravitational field strength on the surface in Nkg⁻¹	9·8	23	3·7	3·7	1·6	11	9·0	270	8·7	8·9
Radius in 1000 km	6·4	70	3·4	2·4	1·7	22	59	700	23	6·0
Mass in 10²⁴ kg	6·0	1900	0·64	0·32	0·07	103	568	2·0	87	4·9

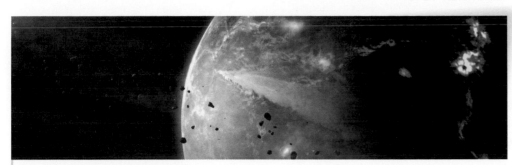

ASTEROIDS AND SLINGSHOTS

NASA's Near Earth Object Programme lists comets and asteroids that have been attracted by the gravitational attraction of nearby planets into orbits that will allow them to pass close to the Earth. Large objects are not likely to hit Earth but smaller objects and dust enter our atmosphere daily where much of it burns up.

Gravity assists

Gravity assists, or **slingshots**, are a method for propelling spacecraft through space. Slingshots can be used to change the direction and the speed of a spacecraft on its voyage through the Solar System. Without this many explorations to parts of our Solar System would not have been successful. Spacecraft have a limit to the amount of fuel carried, so there would be limits on the distance they could travel.

If a planet is considered stationary, an approaching spacecraft on a parabolic trajectory alters its direction but returns to the same speed on departure. However if a spacecraft approaches a planet whose path is in a similar direction to the spacecraft it will accelerate overall. Conversely if the planet approaches in the opposite direction the spacecraft will have decelerated overall.

During 2014, after three Earth gravity assists and a Mars gravity assist, the Rosetta spacecraft rendezvoused with the surface of a comet (Comet 67P/Churyumov Gerasimenko) in an effort to study its composition and structure.

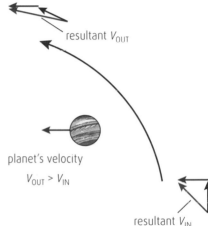

resultant V_{OUT}

planet's velocity

$V_{OUT} > V_{IN}$

resultant V_{IN}

STELLAR FORMATION

In the formation of the planets and stars it is thought that particles combined by electrostatic forces until they were large enough that their **gravitational forces** caused them to have further inelastic collisions resulting in a further increase in mass. Molecular clouds in our galaxy can collapse until a new star is formed as the atoms and molecules fuse together.

Black holes

Black holes are formed when very large dense stars collapse near the end of their life. The collapse creates a **gravitational field** which is so strong that all matter and even light are attracted into the star. The escape velocity within a black hole is theoretically greater than the speed of light so not even light can escape, and the black hole appears as an area of complete darkness when the region is observed.

Black hole

💭 THINGS TO DO AND THINK ABOUT

Cosmology is popular on TV. Try to watch some universe videos by scientists such as Brian Cox and others. Find out what happens to different types of stars at their life-end.

SPECIAL RELATIVITY

In 1905, at the of age 26, Albert Einstein published one of the most important papers in the history of physics, which presented his theory of Special Relativity, showing how to interpret motion in different inertial frames of reference. We learn here some of the consequences.

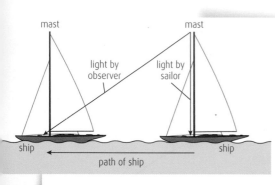

mast   mast

light by observer   light by sailor

ship   ship

path of ship

RELATIVITY AND THE SPEED OF LIGHT

Galilean invariance

Galileo Galilei in 1632 had a **classic view of relativity**. If you watch a sailor drop a rock from the top of a mast, does it land at the base of the mast or some distance back because of the ship's motion?

You might think a distance back but it lands at the base of the mast. The sailor sees the rock fall straight down but you see the rock fall at an angle! The motion of the rock is relative to who is viewing it. Below deck a scientist is measuring the speed of a ball rolling across a table. As the scientist does not know whether the ship is stationary or moving, Galilean invariance says that the **laws of motion** are the **same in all inertial frames**.

Newtonian relativity

1 You sit in your chair and you are at rest. However the Earth is moving and so you are moving?

2 You are walking in a moving train. Is your speed relative to the floor of the train or the ground outside?

Observers in different frames of reference measure events with different results. In **Newtonian mechanics** we have an **absolute space and time** and we can set different frames of reference. Newton's laws and gravity hold.

Einstein's relativity

Einstein built on the work of the Scottish physicist **James Clerk Maxwell** who, 40 years earlier, mathematically proved that light was an electromagnetic wave and in a vacuum exists only at one fixed speed. c = 186 282 miles per second or $c = 299\,792\,458\,\mathrm{ms^{-1}}$.

For ten years, Einstein considered an extremely tall ship version of Galileo's example. What would happen if the rock was a beam of light? The sailor sending the beam of light and you both see the beam land at the base of the mast. To you watching the beam of light, it has travelled a greater distance than the sailor sees. To you, even though its the same event, if the measured distance has increased, the measured time taken must have increased also!

In 1905 he had an insight. **For light to travel at the same speed, you have to change time.** A new principle of relativity! You measured a greater distance and a longer time so both persons get the same result.

Einstein built his theory on two laws:

- The **speed of light** in a vacuum is **the same** for all observers, regardless of their relative frame of reference.

- The **laws of physics** do not change, even for objects moving in **inertial frames**. (inertial = constant speed).

Time and space are no longer absolute. **Einstein** postulated that measurements of both **space and time** for a moving observer are changed relative to those for a stationary observer. These effects are non-intuitive to us and take some time to think about. Einstein had linked the three dimensions of space with the dimension of time to create a **four dimension space-time continuum**.

contd

These **relativistic effects** are only noticeable at speeds approaching the speed of light which is why Newtonian mechanics are good enough to take a rocket to the Moon. However, for Global Positioning Satellites, atomic clocks on board measuring signals travelling to the satellites in space have to adjust their time taken to account for Einstein's General Theory of Relativity or the positioning is wrong.

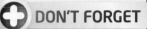

DON'T FORGET

Newton's laws are simple and accurate until speeds of about 0.1 *c*.

TIME DILATION

The fact that the speed of light should not change regardless of the speed of the observer overturned scientific thinking about time. Einstein showed that a stationary observer will measure time more quickly than an observer moving with a relative velocity.

Consider a space ship moving at high speed. An astronaut inside times a beam of light from the floor to a mirror on the ceiling and back. Since the astronaut is in the same frame of reference as the event this time is **time *t***. An astronaut outside sees the beam travel a different longer distance and measures a longer time. That astronaut's clock is running slower and the longer time is **time *t'*** (*t'* is known as *t* **prime**).

The clock of a moving observer appears to be **running slower** and a **longer time** is measured. This is known as **time dilation**. It does not matter whether the rocket is moving and the observer stationary, or the observer is moving and the rocket stationary.

t is the shorter time in the frame of reference as the event. *t'* is the longer time in the relatively moving frame of reference. The relationship between the times is

$$t' = t\frac{1}{\sqrt{1-\frac{v^2}{c^2}}} \quad \text{or} \quad t' = \frac{t}{\sqrt{1-\frac{v^2}{c^2}}}$$

This is sometimes shown as $\gamma = \frac{1}{\sqrt{1-\frac{v^2}{c^2}}}$, where γ is the Lorentz factor (symbol gamma).

γ appears in various relativity equations. Try it with $v = 0.1\,c$, $0.5\,c$ and $0.9\,c$.

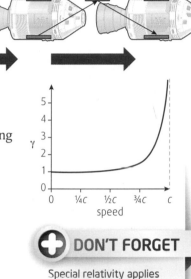

DON'T FORGET

Special relativity applies only when observers are moving at constant velocity.

ONLINE

Follow the link 'Einstein year 1905' at www.brightredbooks.net for more information.

LENGTH CONTRACTION

As objects move through space-time, space as well as time undergoes changes in measurement. The lengths of objects appear to be contracted when they move by us at relativistic speeds. Contraction only takes place in the direction of motion.

l is measured in the same frame of reference as the object. *l'* is the shorter relativistic length.

$$l' = l\sqrt{1-\left(\frac{v}{c}\right)^2}$$

VIDEO LINK

Learn more about relativity by watching the clip at www.brightredbooks.net

THINGS TO DO AND THINK ABOUT

For interest, the equation to calculate the **speed of light** is simply $c = \frac{1}{\sqrt{\varepsilon\mu}}$

ε and μ are values for free space, where ε is the **permittivity** and μ the **permeability** of free space. Find the values and calculate *c*.

As you travel towards the speed of light your **mass** tends towards **infinity**. This is another reason why **nothing** travels faster than the speed of light.

It is usual to consider **relativistic equations** at speeds $> 0.1\,c$.

General relativity applies when objects are changing direction or accelerating. In another paper, published in 1907, Einstein considered a man falling from a roof. His frame of reference is accelerating. He could think he is at rest. Acceleration and gravitation are two aspects of the same force.

ONLINE TEST

Head to www.brightredbooks.net and test yourself on special relativity.

DOPPLER AND HUBBLE

We will examine the Doppler effect and see how it was used by Hubble to explain the movement of galaxies in the universe.

observer 1

observer 2

THE DOPPLER EFFECT FOR SOUND

A stationary siren emits sound of **frequency** $f = \frac{v}{\lambda}$ where v is the **velocity** of sound, $v = 340\,\text{ms}^{-1}$.

When a vehicle with a siren **moves towards** you, the sound has a **higher pitch**, and **shorter wavelength**, than it would have if the vehicle was stationary. As the vehicle passes, you observe a drop in the pitch of the sound and there is an increase in wavelength. This is an example of the **Doppler effect**.

The **source frequency and wavelength do not change**. The **received frequency and wavelength** are due to the **velocity of the siren** relative to the observer.

The apparent or observed frequency can be calculated. $f_o = f_s \left(\frac{v}{v \pm v_s} \right)$

f_o is the observed frequency, f_s is the frequency of the source.

In the denominator of the expression, the $\pm$ indicates whether the source is moving towards or away from the observer.

A negative sign is used when the frequency increases, and the source moves towards the observer.

A positive sign is used when the frequency decreases, and the source moves away from the observer.

The Doppler effect is observed in sound but was originally introduced in 1842 by Christian Doppler to explain the coloured light of binary stars. However the Doppler equation above cannot be used with light as relativistic effects are required.

long wavelength
low frequency

short wavelength
high frequency

Doppler effect

Example:

A siren travels at 20 ms⁻¹ towards you emitting a note of frequency 10 kHz. What is the observed frequency?

Solution:

$$F_0 = F_s \left(\frac{v}{v \pm v_s} \right) = 10\,000 \left(\frac{340}{340 - 20} \right) = 10\,625 \text{ Hz}$$

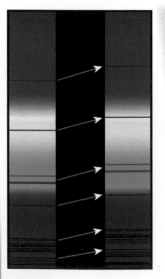

REDSHIFT

We see stars because of the light they emit. Our Sun emits a **continuous spectrum** which we see in a rainbow and adds up to white light. When viewing distant stars there may be a **redshift** if the star is moving rapidly **away** from us or a **blueshift** if the star is moving **towards** us. Some stars in spiral galaxies are moving towards us while others move away.

Black **Fraunhofer lines** are absorption spectra and their pattern can identify that a distant star's atmosphere contains chemicals similar to those in our own Sun. The amount and direction of shift the pattern may have from its normal position is due to the **relative velocity** of the star to us.

Astronomers often measure redshift (z) in terms of the observed (o) and rest (r) wavelength. z has no units. z is **positive** when stars move away, and **negative** when stars move towards us. $z = \frac{\lambda_0 - \lambda_r}{\lambda_r}$

Redshift is the change in wavelength to the emitted wavelength.

Redshift at low relative speeds can also be calculated from: $z = \frac{v}{c}$

contd

Redshift is the ratio of the relative velocity of the star to the speed of light.

Remember that stars and galaxies often have very high velocities which means that astronomers have to make corrections for relativistic effects which are beyond this course.

DON'T FORGET

The **component** of velocity towards us is calculated. The **direction** of the stars can also be found.

HUBBLE

Until the work of Edwin Hubble in the 1920s, astronomers considered the known universe was no further than our Milky Way.

Hubble discovered pulsing stars inside what were thought to be gas clouds and realised these were beyond our galaxy. The pulsations of these stars known as **Cepheid variables** (and sometimes called **standard candles**) allowed Hubble to calculate their **distance from Earth**. Astronomers realised the gas clouds or nebula were **distant galaxies**.

Hubble's studies of these galaxies made him realise that the galaxies exhibited a **redshift**, and hence a **recession velocity**, which was increasing in direct proportion to the distance from us. More studies of galaxies since have added to his data and to its accuracy. They confirm **Hubble's law**: $v = H_0 d$.

The current calculation of Hubble's constant, $H_0 = 2 \cdot 34 \times 10^{-18} s^{-1}$.

Most astronomers use light years and parsecs in measurements, but in our course we stick with the SI units.

Hubble's studies in the **redshift** showed

- that most **galaxies are receding** from us and
- that the **recession velocity increases with distance**.

There are exceptions. The Andromeda galaxy, as the largest galaxy in our Local Group, is instead rushing towards us, and as it gets closer, the gravitational pull will be increasing, and it will collide with our Milky Way in about four billion years.

VIDEO LINK

Learn more by watching the great clips at at www. brightredbooks.net

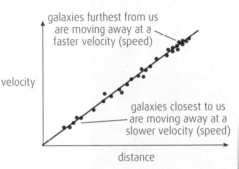

Hubble's Law
velocity = Hubble constant × distance

galaxies furthest from us are moving away at a faster velocity (speed)

velocity

galaxies closest to us are moving away at a slower velocity (speed)

distance

DON'T FORGET

Hubble has a telescope named after him which has given us far-reaching images of the universe.

AGE OF THE UNIVERSE

Hubble's law suggests that most galaxies are receding away from us that space and the universe is expanding. If we assume that this expansion is at a steady rate, we can make an estimate of the age of the universe: reverse the rate of expansion and we can estimate when the stars and galaxies began moving?

Time $t = \frac{d}{v}$ and from Hubble rearranged we have $\frac{d}{v} = \frac{1}{H_0}$. So time $t = \frac{1}{H_0}$

The latest estimates of the age of the universe by reversing the expansion are ~14 billion years.

ONLINE

Follow the link at www.brightredbooks. net and check out the Doppler animation.

THINGS TO DO AND THINK ABOUT

If we know a particular electromagnetic wave was emitted at the star with a wavelength of $\lambda = 486\,nm$ but it was detected here on Earth with a wavelength $\lambda = 520\,nm$, we can say its redshift is

$z = \frac{(520\,nm - 486\,nm)}{486\,nm} \sim 0.07$

Redshift can be caused by the Doppler effect but a cosmological redshift is where the light waves have been stretched by the expansion of space.

ONLINE TEST

Revise your knowledge of Doppler and Hubble by testing yourself at www. brightredbooks.net

THE EXPANDING UNIVERSE

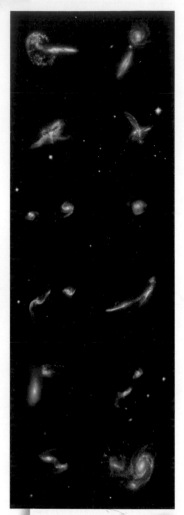

Considering theories for our universe, we see the need for dark matter and dark energy.

THEORIES ON THE EXPANDING UNIVERSE

Is the universe finite or infinite? Has it always existed or did it begin at some point in the past? Can we work out its size or age? These are the types of questions that astronomers have pondered for hundreds if not thousands of years.

Newton, having discovered the law of gravity, realised that it was always an attractive force, and so a finite universe would collapse in on itself. **Olbers** thought that the universe must be finite or the sky would be filled with light everywhere.

Einstein's equations, from his General Theory of Relativity about gravity, told him the universe should either be expanding or collapsing, which contradicted his view that the universe was static.

Hubble, studying the redshift from galaxies, came to the explanation that the universe is expanding. Most scientific studies tend to reinforce this idea.

The expanding universe is **finite** in both **time** and **space**.

If the **universe** has always been **expanding**, will it continue to expand forever, will it expand at a constant rate or will it cease expanding and collapse in again?

It is thought that what happens to the universe can be determined by measuring how fast the universe expands **relative** to how much matter the universe contains. For the last 80 years, astronomers have been making increasingly accurate measurements of

- the **rate** at which the **universe expands**
- the **average density** and **mass** of matter in the universe.

ONLINE

You can find more about "my greatest blunder" at www. brightredbooks.net

MASS OF A GALAXY

We can estimate the **mass** of a galaxy by the **orbital speed** of the stars within it.

Stars in a **more massive galaxy** will **orbit faster** than those in a lower mass galaxy. The **greater gravity force** of the massive galaxy will cause larger **accelerations** of its stars. Star speeds tell us how much gravity there is in the galaxy. As gravity depends on mass and distance, the size of the star orbits allows us to derive the **galaxy's mass**.

We find that although most light and other radiations emit from near the centre, most mass is calculated as being near the edge of a galaxy. What is happening?

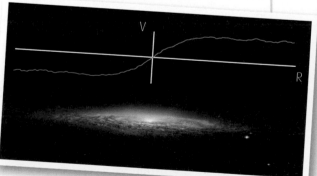

Spiral galaxy

DARK MATTER

The stars and gas in most galaxies are moving much quicker than their luminosity predicts, so something must be adding to gravity that we cannot see. Does a new type of matter exist? Astronomers consider it may be made of planets, brown dwarfs, white dwarfs, black holes, or neutrinos but as it does not seem to emit any radiation we can detect (telescopes do not see it) it is assumed to be made of **a new type of matter or exotic particles** which we do not yet know about.

The overall mass of this new **dark matter** must be about five times the mass of matter we do know about.

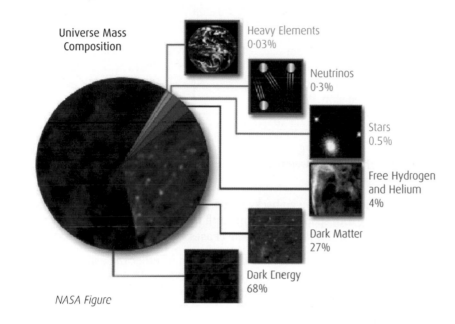

Universe Mass Composition

Heavy Elements
0·03%

Neutrinos
0·3%

Stars
0.5%

Free Hydrogen and Helium
4%

Dark Matter
27%

Dark Energy
68%

NASA Figure

 **DON'T FORGET**

Dark energy is something that overcomes the force of gravity.

DON'T FORGET

Dark matter and dark energy are known by their effect on ordinary matter and energy.

DARK ENERGY

Gravity is the force which slows down the expansion of the universe. However not only does the universe appear to be expanding but the **rate of expansion** appears to be **increasing**.

That acceleration implies an **energy** that acts in **opposition to gravity**, which would cause the expansion to accelerate. Just like dark matter, this energy has never been detected directly through observation, and it has been given the name **dark energy**.

Dark energy is assumed to exist **throughout the universe**. Where galaxies are grouped together gravitational force may predominate but distant galaxies are driven apart by this unknown energy.

THINGS TO DO AND THINK ABOUT

You must remember that good physicists like to question theories. For example, if the universe is expanding, why is the light from distant galaxies not too faint too see, as predicted by the Big Bang theory (which we study next)? Try to follow modern physics questions online. One team reported in 2014 an x-ray spike from galaxies, which may come from dark matter.

The existence and nature of dark matter and dark energy are big questions in physics today. Dark matter and dark energy are thought to make up about 95% of the total content of the universe, so that's 95% we don't know about!

ONLINE

Follow the links at www. brightredbooks.net for more on this topic.

 **VIDEO LINK**

Head to www. brightredbooks.net and watch the clips on dark energy and the accelerating universe.

ONLINE TEST

Head to www. brightredbooks.net and take the test on the expanding universe.

THE BIG BANG THEORY

The theory for expansion that came to be known as the Big Bang is examined along with some of its problems and the evidence for it.

THE THEORY

We have seen Hubble's discovery that shows an expanding universe. Hubble realised that the light from distant galaxies is red-shifted and that the more distant the galaxies are, the greater the red shift. We can interpret this as the universe expanding. The space between clusters of galaxies is expanding and therefore the distance between galaxies is increasing.

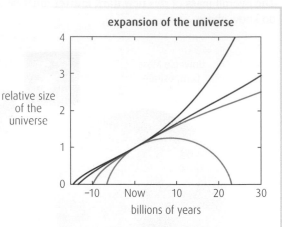

expansion of the universe

If we reverse this theory, and we go back in time, the galaxies and all matter and energy in the universe along with space itself shrink until they reach a point known as a **singularity**. Of course at this point even **time ceases**.

In the expanding universe, the future could depend on whether the expansion continues steadily, accelerates or decreases. No matter which model we follow, if we reverse the timeline, we can see a singularity as a point in time when the universe was created.

Time, **space**, **matter** and **energy** all start at this point. The Big Bang theory, first proposed by **Lemaître**, is that the universe began from a theoretically infinite **density** and **temperature** at a **finite time** in the past. This singularity then **expanded** over billions of years, expanding **space**. From the initial point the universe went through sub-atomic particles, atoms, and gas clouds, which gravity then formed into stars and galaxies.

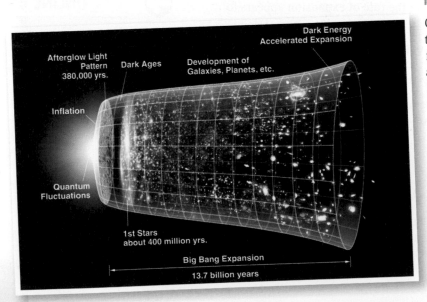

Inflation

One main problem with the initial Big Bang theory was that there was not enough time for the universe to reach thermal equilibrium and, in terms of **energy**, it looks the **same in all directions**. In 1980 an American, Alan Guth, gave an **inflation theory** showing an **extremely rapid expansion** in the **early stage** driven by an energy which may be similar to dark energy.

Beginning in 2001 the Wilkinson Microwave Anisotropy Probe (WMAP) NASA programme provided evidence in support of the theory of the inflationary period.

After the initial expansion the **rate of expansion** is thought to have slowed down due to the pull of **gravity** before **dark energy** appears to be taking over and accelerating the **expansion rate** again.

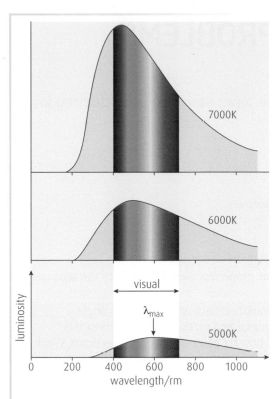

STELLAR OBJECTS

When astronomers see into space, they gather data from all parts of the **electromagnetic spectrum**. Looking at stars, they see the **temperature** of **stellar objects** is related to the distribution of emitted radiation over a wide range of **wavelengths**.

- **Hotter stars** are **bluer**, emitting more light at **short wavelength**, high frequency. **High peak** and **large area** of graph.

- **Cooler stars** are **redder**, emitting more light at **long wavelength**, low frequency. **Low peak** and **small area** of graph.

The wavelength of the peak wavelength of this distribution is shorter for hotter objects than for cooler objects.

The diagram above shows the radiation spectra for a blue star, a white star like our Sun, and a red star.

The shape of these curves is similar to any hot object emitting radiation with intensity rising to a peak at a certain value of wavelength. This curve describes any **black body radiation**.

Hertzsprung-Russell diagrams plot temperature against brightness. Most stars are in the main sequence with our Sun in the yellowed part.

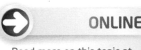

ONLINE

Read more on this topic at www.brightredbooks.net

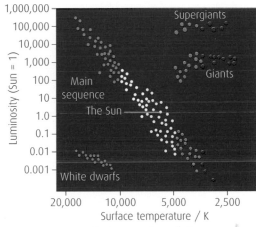

Hertzsprung-Russell diagram

COSMIC MICROWAVE BACKGROUND

The Cosmic Microwave Background (CMB) is radiation from light remaining from around the time of the first formation of atoms. As space expanded the radiation has stretched to become microwaves.

The radiation was predicted as a result of the Big Bang theory and then discovered by accident by two radio astronomers. This radiation is everywhere around us and its temperature is only 3 K. The radiation curve matches that of a black body radiation and is some of the main evidence for the Big Bang theory.

The NASA Cosmic Background Explorer (COBE) maps minute variations in thermal radiation. Contributions from the Milky Way, included in the upper background image, have been eliminated in the lower image to reveal ripples in the thermal radiation left over from the Big Bang.

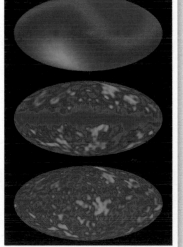

VIDEO LINK

Head to www. brightredbooks.net and watch the clips 'Hubblesite' and 'Big Bang plus'.

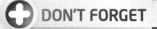

DON'T FORGET

Cosmic Background Radiation (CBR) contains about 99.9% of all the photons in the universe.

THINGS TO DO AND THINK ABOUT

Find out how the WMAP project and the Planck observatory have added to our understanding of the CBR.

ONLINE TEST

Test your knowledge of the Big Bang theory at www. brightredbooks.net

OUR DYNAMIC UNIVERSE PROBLEMS

Practice and revise with help from these examples from the topics you have been studying in this unit.

MOTION AND FORCES

1 Distinguish between distance and displacement.
2 What is velocity?
3 What is a vector quantity?
4 A man walks 300 m east then 200 m south. What is his displacement?
5 What does the gradient of a velocity-time graph represent? What does the area under a velocity-time graph represent?
6 A football is kicked into the air with an initial velocity of 40 ms⁻¹ at an angle of 25° to the ground. Calculate the horizontal and vertical components of this velocity. Calculate the time in flight and how far away the ball touches down. What assumptions have you made?
7 What is acceleration?
8 Describe a method for measuring acceleration.
9 From this velocity-time graph, draw the corresponding acceleration-time graph.

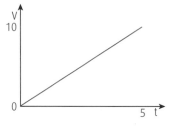

10 Label both graphs from question 9 to describe the motion appropriately.
11 Derive the equations of motion from basic definitions.
12 What is the final speed of a car which accelerates from rest at 5 ms⁻² for a distance of 20 m?
13 Define the Newton.
14 A toy rocket of mass 2 kg takes off with a thrust of 300 N. Calculate the initial acceleration.
15 A skier of mass 80 kg descends 100 m down a slope. If her speed at the bottom is 30 ms⁻¹, what was the average frictional force?

COLLISIONS AND IMPULSE

16 What is momentum?
17 How is the law of conservation of linear momentum applied?
18 What is an elastic collision?
19 What is an inelastic collision?
20 A 6 kg vehicle moving at 2 ms⁻¹ collides elastically with a stationary 3 kg vehicle. What is the new velocity of the stationary vehicle if the first vehicle is slowed to 1 ms⁻¹?
21 A gun of mass 100 kg fires a bullet of mass 1 kg with a velocity of 200 ms⁻¹. Calculate the gun's recoil velocity.
22 A skier goes down a hill then part way up an opposing hill to a stop. What energy equation could describe this motion?

contd

23 Write a relationship for impulse based on its cause.

24 Write a relationship for impulse based on its effect.

25 Write two units for impulse.

26 A 800 kg car crashes into a wall at 20 ms⁻¹. If the contact time was 400 ms calculate the change in momentum of the car and the average force exerted. What is the difference if a softer tyre wall is hit?

GRAVITATION

27 Define gravitational field strength.

28 A cannonball is fired horizontally at 35 ms⁻¹ from a height of 5 m. How long does it take to impact the ground? What is its vertical component of velocity then? How far away does it make impact? What is the angle it impacts from the horizontal?

29 What would be your weight on Mars?

30 What is meant by the inverse square law of gravitation?

31 Calculate the gravitational force due to two pupils of mass 60 kg and 70 kg at a distance of 2 m. What assumptions do you need to make?

DON'T FORGET

A light year is the distance travelled by light in one year. $d = vt$
$= 3 \times 10^8 \times (1 \times 365 \times 24 \times 60 \times 60)$
$= 9.46 \times 10^{15}$ m.

THE UNIVERSE

32 Two space ships travel towards each other, each having a speed of 2×10^8 ms⁻¹, One sends out a beam of light. What speed does each space ship measure the speed of the light to be?

33 An alien in a space craft travels to Earth at a speed of 0·6c. The alien times the journey at 0·5 years. What time does the European Space Agency time the journey from Earth?

34 A rocket has a length of 35·0 m when measured on its launch pad on Earth. When it passes an astronaut on the Moon some time later it is moving at a steady speed of $1·50 \times 10^8$ ms⁻¹. Calculate the length as measured by the astronaut on the Moon.

35 A police car is fitted with a siren emitting a frequency of 1200 Hz. As the car drives past an observer at 30 ms⁻¹, what three frequencies are heard? The speed of sound is 340 ms⁻¹. What is the shortest wavelength observed?

36 What is the name of the effect in Q35?

37 The distance to our sun is $1·44 \times 10^{11}$ m. Convert this distance into light seconds and light years.

38 How many light years are there across our Milky Way galaxy?

39 Light of hydrogen contains a colour of wavelength 434 nm. When the spectral line is found in light from a distant galaxy, it is noted that its observed wavelength is now 466 nm. Calculate the fractional change in wavelength and the redshift z for this galaxy. Calculate the velocity of this galaxy relative to Earth.

40 Calculate the value of the redshift, z, for a star travelling away from Earth at a speed of $2·7 \times 10^7$ ms⁻¹. You can use non-relativistic calculations.

41 Use Hubble's law to calculate the speed a galaxy is moving away from us if it is 1×10^9 light years away.

42 What happens to the value of peak wavelength of black body radiation as an object gets hotter? What happens to the total amount of energy emitted as the temperature of a star increases?

THINGS TO DO AND THINK ABOUT

Check with the previous pages to ensure you can answer the questions correctly and reinforce your learning.

MONITORING AND MEASURING ALTERNATING CURRENT

In a power station, a generator has conductors which move up and down through a magnetic field as they rotate. **Current** and **voltage** are induced in **alternating directions**. A.c. power is then sent to our homes using transformers with **low power loss**. We can **monitor** and **measure a.c.** with an **oscilloscope**. We can measure **peak** and **r.m.s.** values. We can also use a multimeter as an **a.c.** voltmeter and ammeter.

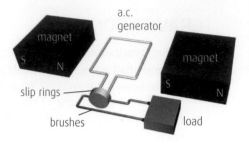

a.c. generator

magnet

slip rings

brushes

load

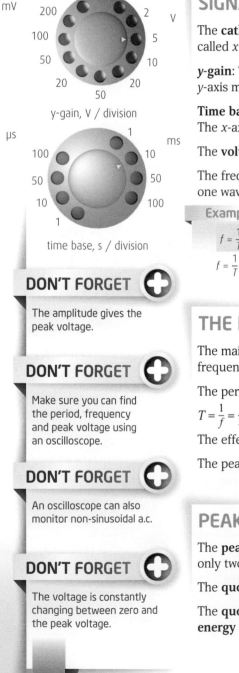

y-gain, V / division

time base, s / division

SIGNALS ON THE OSCILLOSCOPE

The **cathode ray oscilloscope** (CRO) has two axes called x and y – just like in maths.

y-gain: The **y-gain** changes the scale on the y-axis. The y-axis measures **voltage** in $V\,cm^{-1}$ or $V\,div^{-1}$.

Time base: the **x-gain** changes the scale on the x-axis. The x-axis measures **time** in $s\,cm^{-1}$ or $s\,div^{-1}$.

The **voltage** alternates in a **sinusoidal** wave with a frequency f.

The frequency can be calculated from the **period** of time for one wave.

> **Example:**
> $f = \dfrac{1}{T}$ On the screen, the period $T = 5 \times 10 = 50\,ms = 0.05\,s$
>
> $f = \dfrac{1}{T} = \dfrac{1}{0.05} = 20\,Hz$

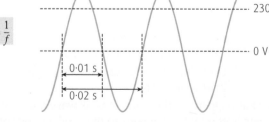

THE MAINS SUPPLY

The mains electricity supply has a frequency $f = 50\,Hz$.

The period of the mains supply: $T = \dfrac{1}{f}$

$T = \dfrac{1}{f} = \dfrac{1}{50} = 0.02\,s$ $T = 0.02\,s$

The effective **voltage = 230 V**

The peak voltage = 325 V approx.

~325 V
230 V

0.01 s

0.02 s

0 V

PEAK AND R.M.S. VALUES

The **peak voltage** (from zero or rest to the peak) is the **maximum** reached by a supply at only two points in each cycle. The peak voltage can be found using an oscilloscope.

The **quoted voltage** is also called the effective or **r.m.s.** voltage.

The **quoted voltage** is the value of an a.c. voltage, which will provide the **same amount** of **energy** as a d.c. voltage.

contd

The relationship between **peak and r.m.s. voltage** is

$$V_{rms} = \frac{V_{peak}}{\sqrt{2}} \text{ or } V_{peak} = \sqrt{2}V_{rms}$$

Current alternates as a sinusoidal wave.

Current has a **peak** and an **effective** or **r.m.s.** value.

The relationship between peak and r.m.s. current is

$$I_{rms} = \frac{I_{peak}}{\sqrt{2}} \text{ or } I_{peak} = \sqrt{2}I_{rms}$$

Examples:

1   The dial on an a.c. power supply shows 10 V. What is the peak voltage?
2   If the peak voltage is 10 V, what is the r.m.s. voltage?
3   An a.c. ammeter reads 5 A. What is the peak current?

Solutions:

1   $V_{peak} = \sqrt{2} \times V_{rms} = 10\sqrt{2} = 14 \cdot 1\,V$

2   $V_{rms} = \frac{V_{peak}}{\sqrt{2}} = \frac{10}{\sqrt{2}} = 7 \cdot 1\,V$

3   $I_{peak} = \sqrt{2} \times I_{Rms} = 5\sqrt{2} = 7 \cdot 1\,A$

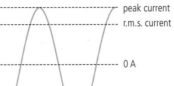

peak voltage
r.m.s. voltage

0 V

peak current
r.m.s. current

0 A

DON'T FORGET

r.m.s. = root mean square, which is from the mathematical proof, but this is not required!

DON'T FORGET

The effective value is always less than the peak value. The value quoted for a.c. power supplies is the effective or r.m.s. value.

DON'T FORGET

a.c. ammeters and voltmeters are calibrated to give the effective or r.m.s. value.

MULTIMETERS

A multimeter can be used as an ammeter, voltmeter and an ohmmeter.

When set to **d.c.** and connected to a battery, the voltmeter range gives a reading of the **steady voltage** and the ammeter range gives a reading of the **steady current**. Calculation of $\frac{V_{dc}}{I_{dc}}$ gives resistance R.

When set to **a.c.** and connected to a laboratory power supply, the voltmeter range gives a reading of the **effective** or **r.m.s.** value of a.c. **voltage** and the ammeter range gives a reading of the **r.m.s.** value of a.c. **current**. Thus in a.c. although the voltage and current values are constantly changing as seen on the oscilloscope, the r.m.s. values are steady. Calculation of $\frac{V_{rms}}{I_{rms}}$ gives resistance R.

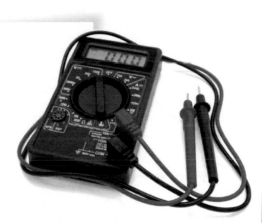

The calculated values of **resistance** can be confirmed using the multimeter as an **ohmmeter**. Simply connect the ohmmeter range across the component (without the component being connected in the circuit) and the value of resistance displayed should be similar to that calculated.

Ohms' law applies for an a.c. waveform:

$$V_{peak} = I_{peak}R$$
$$V_{rms} = I_{rms}R$$

ONLINE

Follow the links at www.brightredbooks.net for more on this.

VIDEO LINK

Watch the clips on a.c./d.c. at www.brightredbooks.net

THINGS TO DO AND THINK ABOUT

1   In practice, measure the peak voltage by measuring the peak–peak voltage, then taking half.
2   r.m.s. = root mean square. This comes from the mathematical derivation of the equations.
3   At any point in an a.c. signal, the instantaneous voltage = the instantaneous current × the resistance, in resistive circuits.

ONLINE TEST

Head to www.brightredbooks.net and test yourself on monitoring and measuring a.c.

CIRCUIT THEORY

In this section, we will consolidate relationships involving potential difference, current, resistance and power. Some you may be familiar with, but calculations may involve several steps.

BASIC CIRCUIT THEORY

Current

Current (*I*) is the **rate** of flow of **charge** (*Q*).

Current is the amount of **charge** in **unit time** (*t*).

Current is measured in **amperes** (**A**).

$$I = \frac{Q}{t}$$
$$Q = It$$

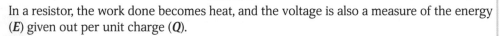

Potential difference

$$V = \frac{W}{Q}$$

The **potential difference** (**p.d.**) between two points is a measure of the **work done** (*W*) in moving **one coulomb of charge** (*Q*) between the two points, such as across a resistor. Potential difference is also called **voltage**.

Voltage (*V*) is measured in **volts** (**V**). $V = \frac{E}{Q}$

In a resistor, the work done becomes heat, and the voltage is also a measure of the energy (*E*) given out per unit charge (*Q*).

Work is done to move charges through the components of a circuit. The work done comes from the energy supplied to the charges as they pass through the source.

Resistance

Resistance (*R*) is measured in **ohms** (Ω).

Ohm's law $R = \frac{V}{I}$

$\frac{V}{I}$ is a constant for most resistors.

Electromotive force

The **electromotive force** (**e.m.f.**) of a source is the **electrical potential energy** supplied to each **coulomb of charge** which passes through the **source**.

E.m.f. is measured in **volts** (JC^{-1}).

Work or energy

Work or energy is measured in **joules** (**J**). $W = QV$ or $E_w = QV$

This can be combined with $Q = It$ to give $E_w = ItV$

Power

Power (*P*) is the **rate** of doing **work**.

Power is the amount of **work done** in **unit time** (*t*).

Power is measured in **watts** (**W**). $P = \frac{E}{t}$

$$P = \frac{E}{t} = \frac{ItV}{t} = IV \qquad P = IV$$

$$P = IV = I(IR) = I^2R \qquad\qquad P = I^2R$$

$$P = IV = \left(\frac{V}{R}\right)V = \frac{V^2}{R} \quad P = \frac{V^2}{R}$$

TOTAL RESISTANCE

Conservation of energy

From conservation of energy, we know that energy is not created or destroyed. The energy supplied per coulomb to the charges as they pass through the source must equal the energy dissipated per coulomb by the charges in the circuit.

The e.m.f. of the supply is equal to the sum of the p.d. round the circuit.

$$E = \Sigma IR \quad \text{or} \quad E = V_1 + V_2 + V_3$$

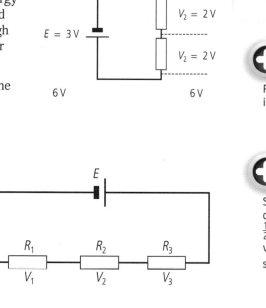

Resistors in series

From conservation of energy:

I is constant

$$E = V_1 + V_2 + V_3$$
$$IR_T = IR_1 + IR_2 + IR_3$$
$$IR_T = I(R_1 + R_2 + R_3)$$
$$R_T = R_1 + R_2 + R_3$$

Resistors in parallel

From conservation of charge:

$$I = I_1 + I_2 + I_3$$

V constant

$$\frac{V}{R_t} = \frac{V}{R_1} + \frac{V}{R_2} + \frac{V}{R_3}$$

$$V\frac{1}{R_t} = V\left(\frac{1}{R_1} + \frac{1}{R_2} + \frac{1}{R_3}\right)$$

$$\frac{1}{R_t} = \frac{1}{R_1} + \frac{1}{R_2} + \frac{1}{R_3}$$

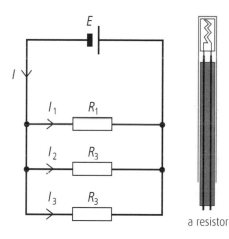

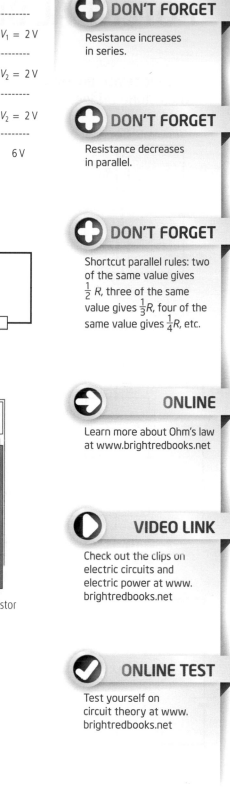

a resistor

Example:

Combine 4Ω, 6Ω and 8Ω resistors.

Solution:

Resistors in series increase the total resistance.
$R_T = R_1 + R_2 = R_3 = 4 + 6 + 8 = 18\Omega$
Resistors in **parallel decrease** the total resistance.
The total resistance is always **smaller** than the **least value** of resistance.

$$\frac{1}{R_T} = \frac{1}{R_1} + \frac{1}{R_2} + \frac{1}{R_3} = \frac{1}{4} + \frac{1}{6} + \frac{1}{8} = \frac{13}{24} \Rightarrow R_T = \frac{24}{13} = 1\cdot8\Omega$$

DON'T FORGET

Resistance increases in series.

DON'T FORGET

Resistance decreases in parallel.

DON'T FORGET

Shortcut parallel rules: two of the same value gives $\frac{1}{2}R$, three of the same value gives $\frac{1}{3}R$, four of the same value gives $\frac{1}{4}R$, etc.

ONLINE

Learn more about Ohm's law at www.brightredbooks.net

VIDEO LINK

Check out the clips on electric circuits and electric power at www.brightredbooks.net

ONLINE TEST

Test yourself on circuit theory at www.brightredbooks.net

THINGS TO DO AND THINK ABOUT

Series: $\quad E = V_1 + V_2 + V_3$

$\qquad\quad I = I_1 = I_2 = I_3$

$\qquad\quad R_T = R_1 + R_2 + R_3$

Parallel: $\quad E = V_1 = V_2 = V_3$

$\qquad\quad\; I = I_1 + I_2 + I_3$

$\qquad\quad\; \frac{1}{R_t} = \frac{1}{R_1} + \frac{1}{R_2} + \frac{1}{R_3}$

POTENTIAL DIVIDER CIRCUITS

Here we revise potential divider circuits and consider how these are applied to the Wheatstone bridge.

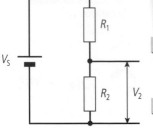

V_S

R_1

R_2 V_2

Potential divider circuit

POTENTIAL DIVIDER CIRCUITS

Potential divider equations: $V_2 = \dfrac{R_2}{R_1 + R_2} V_s$ or $\dfrac{V_1}{V_2} = \dfrac{R_1}{R_2}$

Example:

Consider a potential divider where $R_1 = 4\,\Omega$ and $R_2 = 2\,\Omega$. Calculate V_2 if the supply voltage is 12 V. Recalculate V_2 if a load resistor $R_3 = 2\,\Omega$ is placed across R_2.

Solution:

$$V_2 = \frac{R_2}{R_1 + R_2} V_s = \frac{2}{4 + 2}12 = 4V$$

$$\frac{1}{R_t} = \frac{1}{R_2} + \frac{1}{R_3} = \frac{1}{2} + \frac{1}{2} = 1$$

so $R_T = 1\,\Omega$

$$V_2 = \frac{R_2}{R_1 + R_2} V_s = \frac{2}{4 + 1}12 = 2{\cdot}4V$$

Note how the voltage divides differently when used under load.

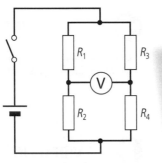

R_1 R_3

V

R_2 R_4

Modern layout

THE WHEATSTONE BRIDGE

The Wheatstone bridge is made of two potential dividers in parallel.

In the Wheatstone bridge, a sensitive voltmeter bridges the outputs of the potential dividers. The Wheatstone bridge circuit is used in two ways.

1   The balanced circuit is used to find the value of an unknown resistor by comparison to standard resistors.
2   The out-of-balance circuit is used to detect small changes in a resistance transducer (thermistor, LDR, etc.).

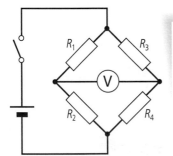

R_1 R_3

V

R_2 R_4

Traditional layout

BALANCED CIRCUIT

When the Wheatstone bridge is balanced, the **potential difference** across the voltmeter is **zero**. The **potential** at each side of the voltmeter is the **same**. **Balance** occurs when the **ratio** of the resistors in each potential divider is the **same**.

When $V = 0$ volts: $\dfrac{R_1}{R_2} = \dfrac{R_3}{R_4}$

R_1 and R_2 are normally fixed-value high-quality resistors.

R_3 is often replaced with a decade resistance box allowing a choice of resistance. R_4 is then an unknown resistor whose value is to be found.

Example:

In a Wheatstone bridge circuit, $R_1 = 100\,\Omega$, $R_2 = 1000\,\Omega$. The voltmeter reads 0 V when the decade resistance box is set at 470 Ω. What is the value of the fourth resistor?

Solution:

$$\frac{R_1}{R_2} = \frac{R_3}{R_4} \quad \frac{100}{1000} = \frac{470}{R_4} \quad R_4 = \frac{470 \times 1000}{100} = 4700\,\Omega$$

An advantage of this circuit is that it does not depend on obtaining accurate readings from ammeters or voltmeters.

Sometimes the voltmeter has a resistor in series with it to protect it from possible large currents in the early adjustments. This is then bypassed to obtain a more sensitive voltmeter.

DON'T FORGET

The balance condition does not depend on the supply voltage.

OUT-OF-BALANCE CIRCUIT

A Wheatstone bridge can be used to measure small changes in resistance.

1   The bridge is initially balanced. The variable resistor R_v is altered until $V = 0\,V$.

2   The value of one of the resistors is changed by a small amount.

3   A voltage is noted on the voltmeter.

4   Increase the change in resistance.

5   A larger voltage appears on the voltmeter.

If the value of one resistor is changed by a small amount in an initially balanced Wheatstone bridge, the out-of-balance p.d. is directly proportional to the change in resistance.

$V \propto \Delta R$

By replacing the $R \pm \Delta R$ resistor block with a resistance transducer, many measurement instruments can be made.

A resistance transducer's resistance will change with a physical property like temperature, light or force.

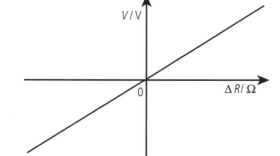

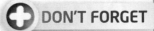

Applications

APPLICATION	RESISTANCE TRANSDUCER	SENSOR
Temperature monitoring	Thermistor or platinum film resistor	head thermistor — rod thermistor — disc thermistor — thermistor circuit symbol; connection to leads; resistance thermometer; connection leads; sheath; insulator
Gas flow meter	Thermistor or platinum film resistor	Gas flow cools the thermometer.
Light meter	LDR (Light-dependent resistor)	
Strain measurement	Strain gauge	

Any changes in the transducer's resistance show up as a change in detector voltage as it varies from zero.

THINGS TO DO AND THINK ABOUT

1   The voltmeter scale will not be left in volts. It will be calibrated to read temperature, light, force or any physical quantity being measured.

2   The voltmeter can be replaced with a computer interface and voltage sensor.

INTERNAL RESISTANCE

A simple cell has an internal resistance and this affects the supply.

INTERNAL RESISTANCE OF A POWER SOURCE

The simple cell is used in many circuits. Most cells show the e.m.f. (the energy supplied per coulomb of charge). But when we use it, the voltage available (called the **terminal potential difference** or t.p.d.) is less than the stated voltage. Why?

In fact, the potential difference across the terminals of a cell, a battery of cells, or any source, decreases as the current drawn from the source increases.

When electrons travel through a source, they have to overcome some resistance. Some of the energy from the e.m.f. is transferred inside the cell by the charges crossing this internal resistance.

An electrical source is equivalent to a source of e.m.f. with a resistor in series, the **internal resistance (r)**.

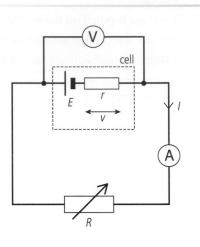

cell has internal resistance

The voltage dropped across the internal resistance of a cell is often called the **lost volts** (v). The voltage available at the terminals is the terminal potential difference or t.p.d. (V).

E = e.m.f. of the source

V = potential difference across the terminals (t.p.d.)

R = resistance of the external circuit

I = current drawn from the source

r = internal resistance of the source

v = the lost volts

Alternative equations

$E = V + v$	the e.m.f. is shared
$E = IR + Ir$	$V = IR$ and $v = Ir$
$V = E - Ir$	t.p.d. = e.m.f – lost volts
$I = \dfrac{E}{R+r}$	current
$r = \dfrac{V}{I} = \dfrac{E-V}{I}$	internal resistance

As circuit resistance R decreases, the current I increases and the lost volts v across the internal resistance increases. Thus the terminal potential difference V decreases.

Example:

A 9 V cell has an internal resistance of 0·5 Ω. What is the t.p.d. when the current drawn is 3 A?

Solution:

$V = E - IR = 9 - (3 \times 0.5) = 7.5 \text{ V}$

DON'T FORGET

The internal resistance r of a cell is considered to be a constant.

DON'T FORGET

The (only) equation given in the data sheet is: $E = V + Ir$

MEASURING E.M.F. AND INTERNAL RESISTANCE

The e.m.f. of a source can be measured using a voltmeter with high resistance as negligible current is drawn to the right.

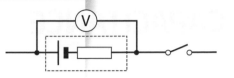

When no current is drawn, there are no lost volts, so the t.p.d. is equal to the e.m.f.

The e.m.f. and internal resistance can be found from the following circuits.

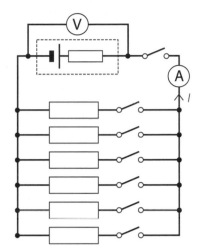

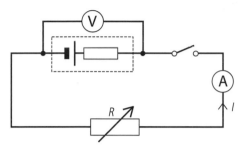

- Measure I and V for various values of load resistance.
- Plot a graph of voltage against current.

When the load resistance is decreased (R reduced or more parallel branches added), the current will increase and t.p.d. will decrease.

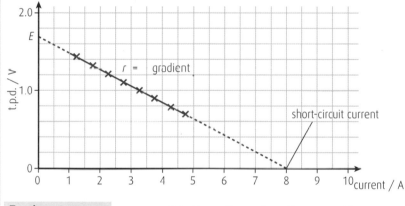

| E = the y-intercept | e.m.f. is the voltage when the current is zero. |
| r = – the gradient | internal resistance from the gradient of the graph. |

The intercept on the current axis gives the maximum **short-circuit current**, which is limited by the internal resistance.

THINGS TO DO AND THINK ABOUT

1. For the mathematical
$E = V + Ir$
$\Rightarrow V = -rI + E$
$y = mx + c \qquad c = E$ and $m = -r$

2. r is constant. When R changes, the ratio $r : R$ changes so the ratio $v : V$ also changes.

3. $V = IR$ should only be used with the external resistance. The whole-circuit equivalent is
$E = I(R + r) \quad$ or $\qquad E = IR + Ir$.

4. Investigate load matching. Maximum power transferred when $R = r$, maximum efficiency when $R > r$.

CAPACITANCE

Learn how capacitors function, investigate the relationship between capacitance, charge and potential difference and find how much energy a capacitor can store.

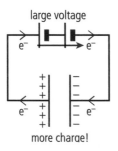

small voltage

charged

large voltage

more charge!

STORING CHARGE

Charge can be stored on parallel metal plates by temporarily connecting them to a direct current (d.c.) source.

Electrons leave one plate and at the same time electrons are added to the other plate.

The energy to cause this transfer of charge from one plate to the other is the work done by the source.

The plates build up charge until the potential difference which is created across the plates is equal in size to the potential difference of the source.

The amount of charge that can be stored depends on the strength or potential difference of the source.

The charge Q on two parallel conducting plates is directly proportional to the p.d. V between the plates.

$$Q \propto V$$

The parallel plates store the energy supplied to them in an electric field between the plates.

When the source is disconnected, the charge and energy are stored.

THE CAPACITOR

A **capacitor** is a component that **stores charge**.

Most parallel-plate designs are rolled into a small cylinder.

To measure the capacitance, the capacitor is switched to the source to charge it (switch c in the diagram). The p.d. V is noted.

The capacitor is then switched to discharge through the coulombmeter (switch d in the diagram), and the amount of charge Q collected is noted.

The experiment is repeated over a range of supply voltages, and results are graphed.

The charge on a capacitor is directly proportional to the p.d. across the capacitor.

Capacitance is the ratio of charge to p.d. $C = \dfrac{Q}{V}$ $Q = CV$

Capacitance is the amount of charge stored per volt.

The unit of capacitance is the **farad (F)**, and one farad is **one coulomb per volt**.

Most capacitors have small values.

microfarad μF 10^{-6} F   nanofarad nF 10^{-9} F   picofarad pF 10^{-12} F

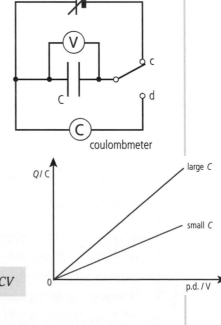

variable d.c. supply

coulombmeter

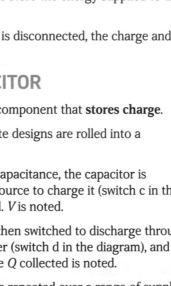

Q/C

large C

small C

p.d. / V

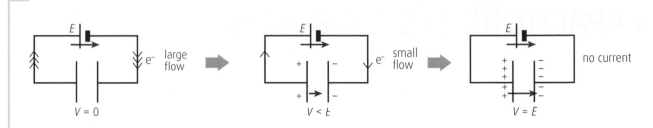

$V = 0$ large flow e^- $V < E$ small flow e^- $V = E$ no current

ENERGY STORED IN A CAPACITOR

Initially, when a capacitor is connected to a source, there is a large flow of electrons off one plate and onto the other.

As charge builds up on the plates (one side negative, one side positive), they create an electrostatic force that opposes the flow. The flow rate reduces.

A potential difference V has been created across the capacitor in opposition to the source's e.m.f. E. However, as E is larger than V, there is still a flow of electrons.

Finally, the potential difference V across the capacitor is as large as the source's e.m.f. E. The capacitor is fully charged. The flow of electrons has reduced to zero.

The cell has done work to charge the capacitor. This work becomes energy stored in the electric field between the plates. (The field pulls on the charges.)

Work done = charge × potential difference

but the potential difference has varied (increased) while charging.

For a capacitor

Energy stored = area under a Q/V graph

$$E = \frac{1}{2}QV = \frac{1}{2}(CV)V = \frac{1}{2}CV^2$$

$$E = \frac{1}{2}QV = \frac{1}{2}Q\left(\frac{Q}{C}\right) = \frac{1}{2}\frac{Q^2}{C}$$

$$E = \frac{1}{2}QV$$
$$E = \frac{1}{2}CV^2$$
$$E = \frac{1}{2}\frac{Q^2}{C}$$

Examples:

1 How much charge is stored in a 5000 μF capacitor when the p.d. across it is
  a 1 V b 5 V?
2 A capacitor takes 880 μC to charge when connected to a 4 V source.
  a What size is the capacitor?
  b How much energy has it stored?
3 How much energy is stored in a 2000 μF capacitor when the p.d. across it is 4 V?
4 A 1000 μF capacitor stores 400 mJ of energy. What charge has been stored?

Solutions:

1 a $Q = CV = 0.005 \times 1 = 0.005\,C$ b $Q = CV = 0.005 \times 5 = 0.025\,C$

2 a $C = \frac{Q}{V} = \frac{880 \times 10^{-6}}{4} = 220\,\mu F$ b $E = \frac{1}{2}QV = \frac{1}{2}880 \times 10^{-6} \times 4 = 1760\,\mu J$

3 $E = \frac{1}{2}CV^2 = \frac{1}{2}2000 \times 10^{-6} \times 4^2 = 1.6 \times 10^{-2}\,J$

4 $E = \frac{1}{2}\frac{Q^2}{C}$ $400 \times 10^{-3} = \frac{1}{2}\frac{Q^2}{21000 \times 10^{-6}}$ $Q = 0.028\,C$

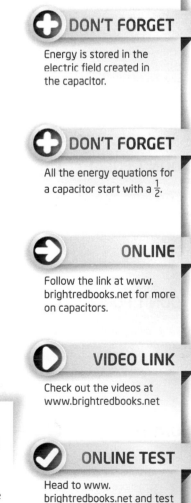

DON'T FORGET

Energy is stored in the electric field created in the capacitor.

DON'T FORGET

All the energy equations for a capacitor start with a $\frac{1}{2}$.

ONLINE

Follow the link at www.brightredbooks.net for more on capacitors.

VIDEO LINK

Check out the videos at www.brightredbooks.net

ONLINE TEST

Head to www.brightredbooks.net and test yourself on this topic.

THINGS TO DO AND THINK ABOUT

1 An insulating material placed between the plates of a parallel plate capacitor is called a **dielectric**. A dielectric increases the capacitance and also allows metal sheets to be rolled without shorting.
2 A capacitor is like a source without internal resistance, as the charges never cross the capacitor. A capacitor can release its energy over a very short time.

CAPACITORS IN D.C. CIRCUITS

Here we learn how capacitors affect current and voltage in d.c. circuits and learn how the charge and discharge time is altered.

DON'T FORGET

When drawing graphs, label with all possible values. Calculate V_{max} and I_{max}. Don't forget the 0 at the origin.

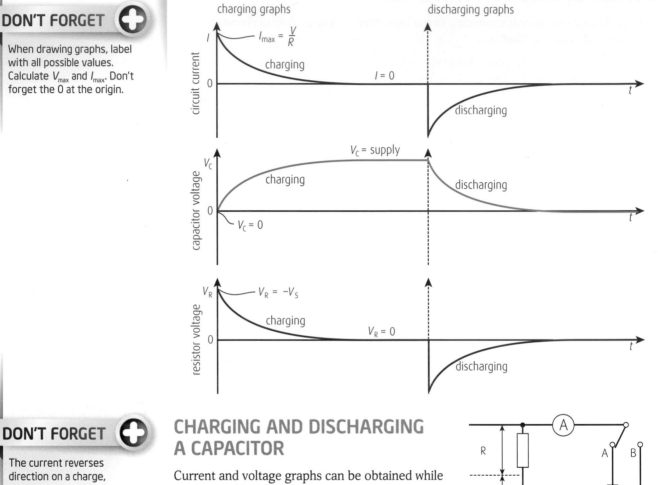

DON'T FORGET

The current reverses direction on a charge, discharge cycle but the capacitor voltage increases then decreases, always in the same direction.

CHARGING AND DISCHARGING A CAPACITOR

Current and voltage graphs can be obtained while charging and discharging a capacitor in an RC circuit by connecting the circuit to a computer and interface. The voltages are sampled, and the software calculates the current using Ohm's law. The value of the resistor used has to be entered. Alternatively, you can obtain readings using voltmeters, ammeter and a stop-clock. Note that meters are not required when the computer is used.

DON'T FORGET

The voltage across the resistor is proportional to the current in the resistor or circuit.

At A: the capacitor is charging through R. At B: the capacitor is discharging across R. The capacitor is charged then discharged.

ONLINE

Follow the link at www.brightredbooks.net to learn more about charge discharge.

CHARGING	DISCHARGING
• The voltage is across the resistor at the start.	• The capacitor voltage is in the same direction as when charging.
• The voltage is across the capacitor when charged.	
• These voltages always add to equal the size of the source voltage: $-V_S = V_C + V_R$	• The current is in the opposite direction as when charging.
• The current is zero after V_C = supply voltage.	

$V = IR$ The current varies as the voltage across the resistor at all points in time.

CHARGING TIME

Increasing the capacitance increases the charging time. More charge will be stored.

$Q = CV$

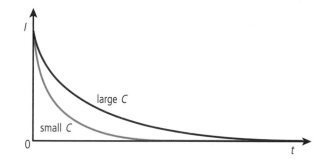

Increasing the resistance increases the charging time. The same charge will be stored.

The current starts smaller but goes on for longer.

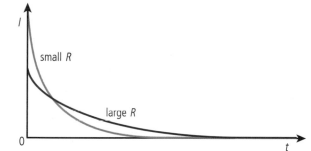

The time to charge or discharge depends on the resistance and the capacitance. The product of $R \times C$ is useful when comparing charging times.

Example:

Find
- **a** the initial current at switch-on
- **b** the voltage of the charged capacitor
- **c** the voltage across the resistor when the voltage across the capacitor is 12V
- **d** the charge stored on the capacitor
- **e** the effect of
  - **i** decreasing R
  - **ii** increasing C
  - **ii** increasing Vs.

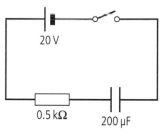

Solution:

a $I_{max} = \dfrac{V}{R} = \dfrac{20}{500} = 0.04\,A$

b $V_C = V_S = 20\,V$

c $V_R = 20 - 12 = 8\,V$

d $Q = CV = 200\mu \times 20 = 4000\,\mu C$

e charging time   **i** decreases   **ii** increases   **iii** stays the same

 THINGS TO DO AND THINK ABOUT

1 The charge/discharge graphs show exponential growth or decay.
2 If the resistor is replaced with a wire when discharging, the time will be shorter and the maximum current will start larger.
3 The circuit with the greatest product of $R \times C$ will take the longest time to charge.

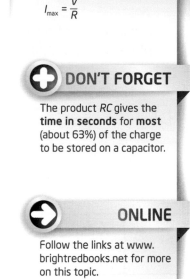

DON'T FORGET

$I_{max} = \dfrac{V}{R}$

DON'T FORGET

The product RC gives the **time in seconds** for **most** (about 63%) of the charge to be stored on a capacitor.

ONLINE

Follow the links at www. brightredbooks.net for more on this topic.

 **VIDEO LINK**

Head to www. brightredbooks.net and check out the clips 'Voltages and time' and 'Time constant'.

 **ONLINE TEST**

Test yourself online at www.brightredbooks.net

CAPACITOR APPLICATIONS

Applications of a capacitor make use of a capacitor's properties to store energy, store charge and block d.c. while passing a.c.

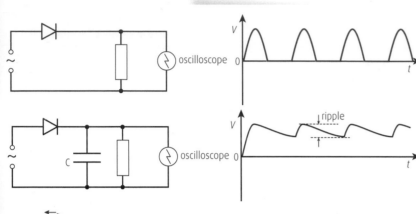

oscilloscope

oscilloscope

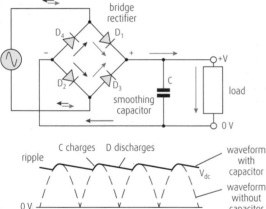

bridge rectifier

D_4 D_1

D_2 D_3

smoothing capacitor

C

load

+V

0 V

C charges   D discharges

ripple

waveform with capacitor

V_{dc}

waveform without capacitor

0 V

resultant output waveform

SMOOTHING RECTIFIED VOLTAGE

An a.c. supply usually supplies its voltage and current in **sine wave** form as the symbol shows in the circuit. Many devices or components require a steady d.c. supply. In the circuit, the component is represented by a resistor.

The simplest way to convert a.c. to d.c. is by **half-wave rectification**. This uses a simple **diode** to block the supply in one direction. This does produce d.c. but the supply arrives in **pulses**. Also only half of the supply is delivered to the component. The oscilloscope shows the voltage across the component.

Addition of a smoothing capacitor in parallel with the component **stores charge** during the pulse, **when the diode conducts**, then **supplies the charge** during the gaps, **when the diode does not conduct**.

The capacitor receives the **same voltage** as the component so when the supply voltage is **increasing** the capacitor voltage is also increasing. When the supply voltage is **decreasing**, the capacitor now starts to top up the supply and continues to supply voltage and also current during the gaps.

Although the capacitor smoothes out the pulses, by acting as a temporary supply itself, it cannot completely replace the normal supply and so a **ripple voltage** is left. The size of this ripple can be reduced by using a larger capacitance, sometimes to almost zero.

A **full-wave rectifier** supplies a constant stream of pulses in one direction. A capacitor across the load smoothes out the ripples in a similar manner as for half-wave rectification. When the supply voltage is increasing the capacitor charges and when the supply decreases the capacitor supplies charge to the load.

ONLINE

Learn more about this by following the links at www.brightredbooks.net

DON'T FORGET

Remember to check out the maximum working voltage as well as the value of capacitance before use.

FILTERS

Filters for blocking d.c.

In some circuits (such as in a radio), you find that a d.c. signal has been added to an a.c. signal.

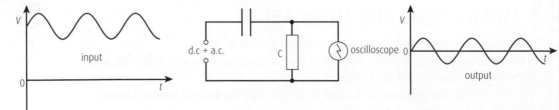

input

d.c + a.c.

C

oscilloscope

output

The capacitor blocks the d.c. components and passes the a.c. component.

contd

Filters for crossover networks in loudspeakers

The capacitor passes high frequencies but blocks low frequencies to prevent damage to the high-frequency speaker.

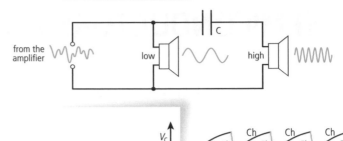

STROBE FLASH LAMP

The strobe-lamp circuit uses a capacitor to store **charge** and **energy** for sudden release across a **neon bulb**. The capacitor charges through a **resistor**. Initially the voltage is across the resistor, but the capacitor's share increases until it is large enough for the neon bulb (which needs a high voltage) to conduct. After the neon has flashed, the process can repeat.

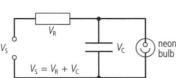

$$V_S = V_R + V_C$$

Ch = charging
Fl = flash

Example voltages

Say the neon bulb in a strobe lamp requires 100V to conduct. A supply of 120V is being used.

- Initially, all the voltage of 120V is across the resistor and maximum current exists.
- The capacitor starts with 0V.
- The capacitor charges and its voltage increases.
- The resistor voltage and the current are decreasing.
- The voltage rises till the capacitor and the neon bulb reach the required striking voltage of 100V.
- Although the supply voltage has not been reached, the neon bulb will strike and the capacitor will discharge rapidly through the bulb.
- When the voltage drops to about 80V the neon bulb cannot continue to conduct and the capacitor will resume charging. The process repeats.

Camera flash

Note that a simpler photographic flash will need a switch in series with the bulb to discharge the capacitor into the bulb.

CAPACITIVE TOUCH SCREENS

Thin layers on the front of a touch screen act as a capacitor. In a **capacitive screen**, you do not need to press hard as charges in your skin affect the charge and electric field and change the capacitance at the point when your finger just touches the screen. A grid can be used to detect your fingers touching several points on the screen at a time.

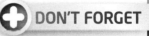

 **THINGS TO DO AND THINK ABOUT**

See how lightly you can operate a phone or laptop capacitive touch screen. (If it is a resistive touch screen you may have to press harder.)

 DON'T FORGET

The time between flashes varies with the values of R and C.

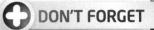 **DON'T FORGET**

The time to the first flash is not the same as the time between flashes.

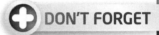 **DON'T FORGET**

The technology of touch screens is rapidly developing and it is worth learning various methods online.

 **DON'T FORGET**

How capacitors pass a.c. can be studied in your next physics course.

 **VIDEO LINK**

Head to www. brightredbooks.net and watch the clips for more.

 **ONLINE TEST**

Revise your knowledge by testing yourself at www. brightredbooks.net

SEMICONDUCTORS AND ENERGY BANDS

In this section, we look at the theory of energy levels in atoms and energy bands in solids.

ENERGY BAND MODELS

From atoms to solids

Looking at **individual atoms**, we see the electrons are contained in **discrete energy levels**. Electrons can only **exist** at these energy levels.

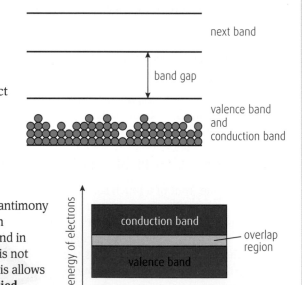

Atom => energy levels

In **solids**, atomic energy levels change to a set of very closely spaced energy levels which blend together into an **energy band**. There are many energy levels within each band due to **thermal** conditions and the **effects of neighbouring atoms**. Bands show the available energies for electrons in the materials. Electrons fill the **bands** closest to the nuclei first and cannot occupy the **energy gaps** between the bands. In the diagram, the dotted line shows the level where the electrons may be found.

The band that the **outer electrons** normally occupy is known as the **valence band**.

All electrons in the **conduction band** are **free electrons**.

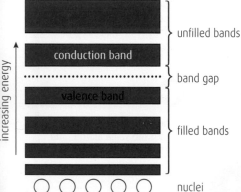

Solid => energy bands.

CONDUCTORS AND INSULATORS

Solids can be categorised into conductors, semiconductors or insulators by their ability to conduct electricity. How the electrons occupy the energy bands tells us whether a material is an **electrical conductor** or **insulator**.

Conductors

Metals and semi-metals (graphite, antimony and arsenic) have **free electrons**. In metals these free electrons are found in the **highest occupied band** which is not completely full of electrons, and this allows the electrons to move with an **applied electric field** and therefore conduct. This band is the **conduction band**.

There are two models you may see for conductors: For some materials, the **conduction band is not full** and the electrons are free to move across in energy levels within this band (with a further unfilled band above). In other materials, the unfilled or conduction band above **overlaps** with the valance band requiring little extra energy to reach its levels.

The electrons in a conductor can easily move into free energy levels and across to neighbouring atoms.

contd

Insulators

In materials such as plastic, rubber, glass and wood, the electrons are all bonded and there are **no** or **few free electrons**.

In an **insulator**, the highest occupied band (the **valence band**) is **full**. The first unfilled band above the valence band is the conduction band. For an insulator, the **gap** between the valence band and the conduction band is very **large**. The addition of heat by normal or room temperatures **cannot** supply enough energy to move electrons from the valence band into the conduction band. There is **no electrical conduction** in an insulator as electrons do not reach the conduction band where they would be able to contribute to conduction.

A large input of energy is more likely to break down the insulator material.

Insulator

DON'T FORGET

An increase in temperature increases the conductivity, and decreases the resistance, of a semiconductor.

ONLINE

Learn more about energy bands by following the links at www.brightredbooks.net

VIDEO LINK

Watch the videos at www. brightredbooks.net for more on semiconductors energy bands.

SEMICONDUCTORS

Silicon, germanium, selenium and some compounds have a **resistance** between good conductors and good insulators.

In a **semiconductor material**, the **gap** between the **valence band** and **conduction band** is very **small** and a **few electrons** can be **excited** from the **filled valence bands** into the **unfilled conduction bands** just by **thermal** vibration. **Electrical conductivity** is between those of metals and those of insulators. Some conduction can take place.

Semiconductor

Negative temperature coefficient

An **increase in temperature increases** the **conductivity** of a **semiconductor** and **decreases** the **resistance**. This is known as a **negative temperature coefficient**. In a normal resistor an increase in temperature decreases resistance due to increased atomic vibrations making it harder for electrons to flow.

Pure semiconductors (intrinsic semiconductors) made of **one material** such as pure **silicon** or **germanium** are **insulators** at very low temperatures.

At **higher temperatures**, when **electrons** in the conduction band move in one direction under the action of an **electric field** from an applied voltage, the **holes** in the valence band will also appear to move in the opposite direction. This occurs as neighbouring electrons in the valance band will also move into the holes, leaving further holes behind, and the process continues.

These materials can change from being an insulator to a conductor and have been used by physicists to lay the foundations of modern electronics.

0 K (no electrons in conduction band)   300 K

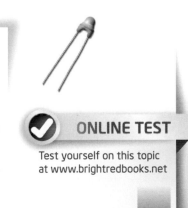

At **zero Kelvin**, the lower band is completely filled with electrons and labelled as the valence band. The upper band is empty and labelled as the conduction band. The **increase in temperature** creates a few electrons in the conduction band and leaves a few holes in the valence band.

THINGS TO DO AND THINK ABOUT

Space five small sheets of paper across a shelf. Each sheet has a potential energy relative to the floor. Now push these sheets together to the same place on the shelf. You will see them slide together to stack up as they cannot be in the same place although they are all on the one shelf. When energy levels of the same value from different atoms come together they form a similar stack or band. All the sheets could be taken to a higher shelf or **energy band**.

A **thermistor** is made with semiconductor material. It has a negative temperature coefficient. Investigate how its **conductivity increases** as **temperature increases**.

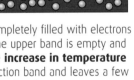

ONLINE TEST

Test yourself on this topic at www.brightredbooks.net

DIODES

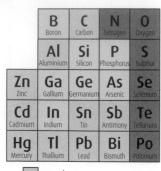

B — Boron
C — Carbon
N — Nitrogen
O — Oxygen
Al — Aluminium
Si — Silicon
P — Phosphorus
S — Sulphur
Zn — Zinc
Ga — Gallium
Ge — Germanium
As — Arsenic
Se — Selenium
Cd — Cadmium
In — Indium
Sn — Tin
Sb — Antimony
Te — Tellurium
Hg — Mercury
Tl — Thallium
Pb — Lead
Bi — Bismuth
Po — Polonium

☐ conductors
☐ semiconductors
☐ insulators

Doping of semiconductors is used to improve semiconductor conductivity and creates n-type and p-type materials. Junctions of n-type and p-type semiconductors create **diodes**.

DOPED SEMICONDUCTORS

Intrinsic semiconductors

Pure (or **intrinsic**) **semiconductors** made of **pure silicon** or **germanium** have four valence electrons in the outer shell, which bond with adjacent atoms so that there are no spare charge carriers at 0 K temperature.

The addition of **heat**, **light** or a **voltage** causes a few electrons to escape from the valence band across the narrow gap into the conduction band, leaving behind **holes**. These can allow a small current to exist so that the **resistance** is **decreased**.

If an electron has gained energy and conducted in one direction a **current** exists. Another electron from a neighbouring atom can now move into the hole left behind, so the hole appears to move along in the opposite direction to the electron direction. The movement of the hole can double the current.

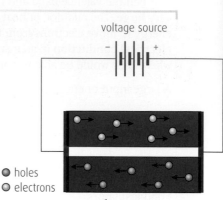

voltage source

○ holes
○ electrons

applied electric field

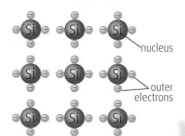

silicon as an insulator

nucleus

outer electrons

IMPURITIES FOR DOPING

Doping a semiconductor (extrinsic semiconductor) is the addition of a small amount of impurity atoms to a pure semiconductor to increase the electrical conductivity of the semiconductor, decreasing its resistance. Although there are a few conducting electrons due to temperature, doping even at extremely low levels greatly increases conductivity.

n-type semiconductor

When a pure semiconductor with four valence electrons is doped with an impurity with five valence electrons, four of these valence electrons form bonds and one electron is left as a free charge carrier. The energy of the fifth electron lies just below the conduction band.

The majority of the free charge carriers are negative, so this is an **n-type** material. As the pure semiconductor and the impurity were both electrically neutral, the n-type material is also electrically neutral.

extra 'free' electron

arsenic (5 electrons)

silicon as an n-type semiconductor

p-type semiconductor

When the pure semiconductor with four valence electrons is doped with an impurity with three valence electrons, three bonds are formed and there is a hole, or missing electron.

This hole can move through the lattice structure, effectively carrying positive charge. The electron energy of a neighbouring atom which is captured by a hole is found just above the valence band.

This is now known as a **p-type** material with a majority of free holes or positive charge carriers.

A hole moves when an electron jumps:

Before   e e e → o e e

After   e e o e e   What has moved?

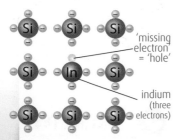

'missing electron' = 'hole'

indium (three electrons)

silicon as a p-type semiconductor

THE JUNCTION DIODE

A **diode** consists of a p-type semiconductor in contact with an n-type semiconductor.

Electrons drift from n-type to p-type to fill adjacent holes. This creates **a depletion layer** at the junction with no free charge carriers. The p-side of the junction gets a small negative charge and the n-side of the junction gets a small positive charge creating a small voltage barrier ΔV against the further drift of charge.

The depletion layer is an insulator.

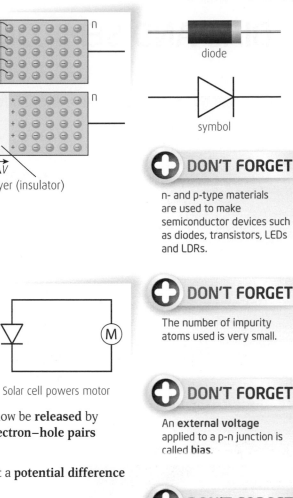

depletion layer (insulator)

diode

symbol

DON'T FORGET

n- and p-type materials are used to make semiconductor devices such as diodes, transistors, LEDs and LDRs.

THE PHOTODIODE

A **photodiode** is a device in which **positive** and **negative charges** are produced by the action of **light** on a **p-n junction**.

When a **photon** of light enters the **depletion layer**, its **energy** can be **absorbed** and an **electron–hole** pair created. Electrons, which had combined with holes to create the depletion layer, can now be **released** by photons of light. This is the **photovoltaic** effect. The number of **electron–hole pairs** varies with the number of **photons**.

Solar cells comprise arrays of many p-n junctions, designed so that a **potential difference** is produced when photons enter the layers.

Solar cell powers motor

DON'T FORGET

The number of impurity atoms used is very small.

Photovoltaic mode

In the **photovoltaic mode**, there is **zero** bias voltage applied, so the photodiode acts as a **solar cell**.

In **photovoltaic mode**, a photodiode may be used to **supply power** to a load.

Photons which are incident on the depletion layer have their **energy absorbed**, **freeing electrons** and **creating electron–hole pairs**. These create a **voltage** or **e.m.f.**

More light creates more **electron–hole pairs**.
- The voltage or e.m.f. is directly proportional to the light irradiance.
- Changes in irradiance produce rapid changes in voltage.
  e.m.f. α irradiance

Each photon releasing an electron-hole pair

DON'T FORGET

An **external voltage** applied to a p-n junction is called **bias**.

DON'T FORGET

When p-type and n-type materials are joined, a layer is formed at the junction. The electrical properties of this layer are used in a number of devices.

ONLINE

Follow this link at www.brightredbooks.net to learn more about the p-n band model.

THINGS TO DO AND THINK ABOUT

In terms of energy bands, the photons increase the number of free charge carriers in the bands.

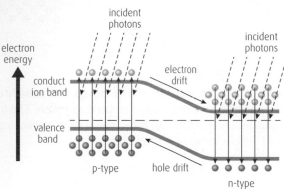

VIDEO LINK

Check out the clips 'The p-n junction solar cells' and 'Types of semiconductor' at www.brightredbooks.net

ONLINE TEST

Head to www.brightredbooks.net and test yourself on junction diodes.

BIAS AND SEMICONDUCTOR DEVICES

Connecting a p-n junction into a circuit affects its operation so it can be used to make semiconductor devices.

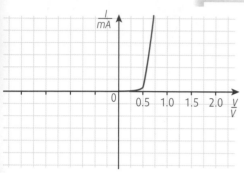

Forward bias: Current when p.d. > 0·6 V, the diode conducts

FORWARD BIAS

When a **voltage** is applied across a p-n junction **in the direction shown** the diode is said to be **forward biased**.

- The **negative** terminal is connected to the **n**-type, **positive** to **p**-type.
- **Electrons** flow from the n-type into the depletion layer, and then into the p-type.
- **Holes** flow from the p-type into the depletion layer, and then into the n-type.
- The **depletion layer** is **reduced.**
- The diode **conducts.**
- Only a very small potential difference (of about 0.6 V for silicon diodes) is needed to overcome the voltage barrier ΔV and the **diode conducts** allowing a large current.

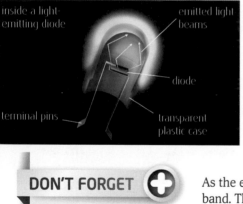

inside a light-emitting diode

emitted light beams

diode

terminal pins

transparent plastic case

THE LIGHT EMITTING DIODE

Light emitting diodes (**LEDs**) are **p-n junctions** which **emit photons** when a current is passed through the junction. The LED operates when **forward biased**.

When the **LED conducts**, the emission of light is caused by **electrons combining** with **holes** to give out **energy** as **photons** of **light**. Each **electron–hole** recombination gives **one photon** out.

Electrons and holes flow in opposite directions. In the **junction region** of a **forward-biased** p-n junction diode, **positive** and **negative charge carriers** can **recombine** to give **quanta** of **radiation**.

As the electrons combine with holes, they fall out of the conduction band into the valence band. The energy emitted depends on the energy difference across the band gap.

Producing colour

To create **photons** with the desired **frequency** or colour, **semiconductor compounds** are used instead of silicon.

Infrared: gallium arsenide (GaAs)

Red: gallium arsenide phosphide (GaAsP)

Green: gallium phosphide (GaP)

Blue: indium gallium nitride (InGaN)

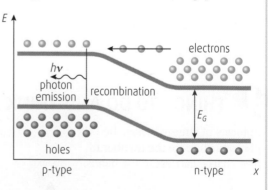

Modern **white** LEDs are based on indium gallium nitride (InGaN). Although this initially emits blue, some blue is absorbed by a phosphor coating and re-emitted as yellow. The blue and yellow combination produces white. There are other methods of producing white. You can research the recent rapid development of white LEDs.

DON'T FORGET

A standard diode conducts in one direction and does not emit light.

DON'T FORGET

Solar cells absorb light, LEDs emit light, the opposite effect.

DON'T FORGET

The LED symbol arrows show the emission of photons.

The LED symbol

LED CHARACTERISTICS

LEDs have to be designed so that the **photons emitted** can leave the surface.

Thin layers are grown and the electrical contacts are sited at the sides of the devices.

The semiconductor material and contacts are encased in plastic.

Advantages of LEDs

- Very low voltage and current are enough to drive the LED.
- Voltage range: 1 to 2 volts.
- Current: 5 to 20 milliamperes.
- Total power output: less than 150 milliwatts.
- Response time: very low, only about 10 nanoseconds.
- LEDs do not need any heating and warm-up time.
- Miniature in size and so lightweight.
- Have a rugged construction and so can withstand shock and vibrations.
- Life span of more than 20 years.

Disadvantages of LEDs

- A slight excess in voltage or current can damage the device.
- The temperature depends on the output power and wavelength.

The device is also known to have a much wider bandwidth compared to the laser. This could be an advantage or disadvantage depending on your use.

DON'T FORGET

In photoconductive mode, the circuit uses a battery in reverse bias to create the wide depletion layer. The current depends only on the light.

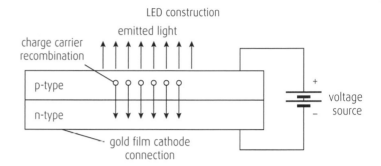

LED construction

REVERSE BIAS PHOTODIODE

When a **voltage** is applied across a p-n junction **in the reverse direction shown** the diode is said to be **reverse biased**.

The **n**egative terminal is connected to the **p**-type, **p**ositive to **n**-type, so the **depletion layer** is **widened** and the diode does not normally **conduct**.

Photons of light shining onto the depletion layer **create electron–hole pairs**. The free charge carriers released lower the resistance and create a **small current**.

The greater the **irradiance** of light, the greater the number of photons, the more free charge carriers and the greater the current.

The current does not depend on the supply voltage. The current is directly proportional to the light irradiance.

Changes in irradiance produce **rapid** changes in current.

The photodiode is said to be in **photoconductive mode**.

DON'T FORGET

The reverse bias photodiode is not specified for the Higher course but has been included for your interest and completeness.

ONLINE

Check out the link at www.brightredbooks.net for more on LEDs.

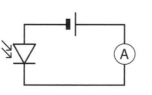

current α irradiance

VIDEO LINK

Watch the p-n junction video at www.brightredbooks.net

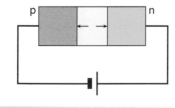
THINGS TO DO AND THINK ABOUT

A p-n diode is grown as a crystal with impurities grown in. The depletion layer is about 10^{-6} m thick.

Investigate the switch-on voltage of different coloured LEDs.

ONLINE TEST

Revise your knowledge of bias and semiconductor devices at www.brightredbooks.net

ELECTRICITY PROBLEMS

Practice and revise with help from these examples from the topics you have been studying in this unit.

MEASURING A.C.

1 Describe how to measure frequency using an oscilloscope.

2 A power pack indicates $7\,V_{rms}$. Calculate the peak voltage.

3 The peak value of an a.c. current is 5 A. Calculate the r.m.s. current value.

CIRCUIT THEORY

4 If a battery supplies 18 J of energy to 2 C of charge, what is its potential difference?

5 If a battery supplies a current of 3 A for 3 minutes, how much charge has been transferred?

6 There are 12 V across a 3 Ω resistor. How much power does the resistor deliver?

7 Derive the expression for the total resistance of any number of resistors in series, by consideration of the conservation of energy.

8 Derive the expression for the total resistance of any number of resistors in parallel by consideration of the conservation of charge.

POTENTIAL DIVIDER CIRCUITS

9 A potential divider circuit has 3 Ω and 6 Ω resistors in series. What is the voltage across each if the supply = 36 V?

10 In a Wheatstone bridge circuit, if R_1 = 2 Ω, R_2 = 4 Ω and R_3 = 6 Ω, what is the value of R_4?

11 What happens for an initially balanced Wheatstone bridge, as the value of one resistor is changed by a small amount?

INTERNAL RESISTANCE

12 Define the e.m.f. of a source.

13 State what an electrical source is equivalent to.

14 Describe the principles of a method of measuring the e.m.f. and the internal resistance of a source.

15 Explain why the e.m.f. of a source is equal to the open circuit p.d. across the terminals of the source.

CAPACITANCE

16 State the relationship between charge and p.d. between two parallel conducting plates.

17 Define capacitance.

18 Define the farad.

19 If 3 V puts 21 000 μC of charge onto a capacitor, what is its capacitance?

20 Explain why work must be done to charge a capacitor.

21 How can you find the work done to charge a capacitor from a Q-V graph?

22 Give three formulae for the energy stored on a capacitor.

23 A 440 pF capacitor is used with a 10 V supply. How much charge and energy are stored on the capacitor?

24 Draw qualitative graphs of current against time and of p.d. against time for the charge and discharge of a capacitor in a d.c. circuit containing a resistor and capacitor in series.

25 A 10 000 μF capacitor is charged through a 500 Ω resistor by a 9 V supply. Find the maximum current in the circuit and the maximum voltage across the capacitor. When does each occur?

26 List possible applications of a capacitor.

SEMICONDUCTORS

27 Where can electrons be found in an atom?

28 Where can electrons be found in a solid?

29 Where are the conduction and valence bands in a conductor?

30 Where are the conduction and valence bands in an insulator?

31 Describe what you know about the energy gap for a semiconductor.

32 State the difference between an intrinsic and an extrinsic semiconductor.

33 How many valence electrons are there in the impurity atom of an n-type and p-type material?

34 State what the free charge carriers in semiconductors are.

35 Describe what a photon does in a solar cells depletion layer.

36 State what forward bias does to a p-n junction.

37 State what electron-hole pairs do in an LED.

38 State the type of materials that make LEDs.

THINGS TO DO AND THINK ABOUT

Check with the previous pages to ensure you can answer these correctly and reinforce your learning.

THE STANDARD MODEL

In this section, we will be learning the scale of everything and introducing the standard model of particle physics.

ORDERS OF MAGNITUDE

The **order of magnitude (oom)** of a number is the number of **powers of ten** indicated.

The speed of light is 3×10^8 ms^{-1}, (oom = 8). The speed of sound is 3×10^2 ms^{-1}, (oom = 2). Light travels 10^6 times (or six orders of magnitude) faster than sound.

To appreciate how important the power of 10 is, consider how a change of power by one **multiplies** a value 10 times.

Draw 10^0 mm
  (or 1 mm) _

Draw 10^1 mm _____

Draw 10^2 mm _____

Draw 10^3 mm _____!

10^3 mm is 1 m, so we can't keep drawing it here. Imagine you are drawing lines continuing to increase by one power of 10 each time. How many more times can you draw a line before it cannot fit the width of Scotland, or in fact, the diameter of Earth ~10^7 m?

The orders of magnitude of length that physicists consider ranges from the very small sub-nuclear particles being discovered inside atoms, to the very large distances to furthest known celestial objects we have observed in astronomy.

10^{-35} m	Planck length	10^1 m	Length of a laboratory
10^{-24} m	Neutrino length	10^2 m	Length of a football pitch
10^{-18} m	Size of an electron Size of a quark	10^3 m	Length of a street
10^{-15} m	Size of a proton	10^5 m	Height of the atmosphere
10^{-14} m	Atomic nucleus	10^6 m	Length of Great Britain
10^{-10} m	Atom	10^7 m	Diameter of Earth
10^{-7} m	Wavelength of visible light Virus	10^9 m	Moon's orbit around the Earth, The farthest any person has travelled. Diameter of the Sun.
10^{-5} m	Red blood cell	10^{16} m	4·24 light years. Distance to Proxima Centauri, next closest star
10^{-4} m	Width of a human hair Grain of salt	10^{21} m	Diameter of our galaxy
10^{-3} m	Width of a credit card	10^{23} m	Distance to the Andromeda galaxy
10^{-2} m	Diameter of a button	10^{27} m	Distance to the next galaxy cluster
10^{-1} m	Diameter of a DVD	10^{29} m	Distance to the edge of the observable universe
10^0 m	Height of a door handle		

THE PARTICLE MODEL

Atoms were first discovered to contain **sub-atomic particles** about 100 years ago. The first to be discovered were the **electron**, **proton** and **neutron**. The atoms of all elements were all considered to be made of these three particles and the Periodic Table to be constructed using the numbers of protons and neutrons.

Rutherford discovered the **nucleus** by firing alpha particles at atoms in 1911. His technique of **collisions** and **subsequent scattering** is used in **particle accelerators** to discover further **sub-nuclear particles**.

THE LARGE HADRON COLLIDER

The 27-kilometre ring of the Large Hadron Collider (LHC) is the **world's largest particle accelerator**. Two high-energy particle beams travel in opposite directions at energies approaching the speed of light before they are made to collide.

The huge ATLAS and CMS detectors shown record the **paths**, **momentum**, and **energy** of the new sub-nuclear particles, allowing them to be individually identified.

Evidence from particle accelerators showed that hundreds of **sub-nuclear particles** could be created. Some have also been detected in cosmic rays. The **particles** are grouped into a classification known as the **Standard Model**.

Large Hadron Collider

The CMS detector

The ATLAS detector

Particle tracks in the LHC

ANTIPARTICLES AND ANTIMATTER

In 1928, the theoretical physicist Paul **Dirac** produced an equation designed to explain the behaviour of the relativistically moving **electron**. The solution to this equation also predicted that for every matter particle there can be another particle, similar and with the **same mass**, but with a **range** of **opposite properties**. These **antiparticles** can be joined together to form **anti-matter**, just as **particles** join to form **matter**.

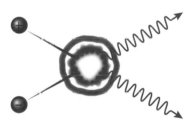

particle-antiparticle annihilation

For the **negative electron**, the antiparticle is the **positron** (from positive electron).

For the **positive proton** there is the **negative anti-proton**.

When a particle and its antiparticle meet they **annihilate** each other. Their mass and kinetic energy are converted into two photons. Momentum is conserved.

Evidence for the existence of antimatter

Antimatter can be created in particle accelerators, but physicist don't know why there is very little in the universe.

THINGS TO DO AND THINK ABOUT

1   You may like to investigate orders of magnitude in relation to **time** and also **mass**.
2   In 1897 J.J. Thomson discovered the **electron**.
3   Rutherford discovered the **proton** in 1918 and Chadwick the **neutron** in 1932.
4   The average **nucleus** diameter is about 10^5 times smaller than that of the **atom**.

DON'T FORGET

Energy can create a particle-antiparticle pair.

DON'T FORGET

A bar above the symbol indicates an antiparticle. Sometimes just the opposite charge is used. proton p⁺, anti-proton p̄⁻, electron e⁻, positron ē⁺.

ONLINE

For more information on powers of 10 and CERN, follow the link at www.brightredbooks.net

VIDEO LINK

Head to www.brightredbooks.net to watch clips on powers of 10 and on antimatter.

ONLINE TEST

Test yourself on the standard model at www.brightredbooks.net

FUNDAMENTAL PARTICLES

Sub-atomic and sub-nuclear particles are now considered **fundamental particles**.

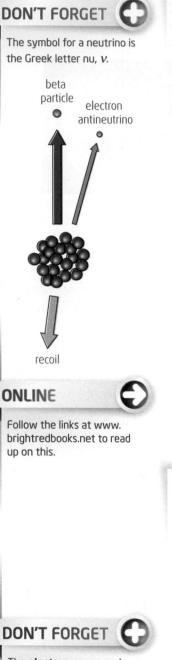

beta particle

electron antineutrino

recoil

EVIDENCE FROM BETA DECAY

A **fundamental particle** is one with no internal structure and is not made of anything smaller. **Atoms** were once considered fundamental, then **electrons**, **protons** and **neutrons**. However **nuclear radiations** probing into and coming from this sub-atomic structure brought up a problem.

Alpha particles are emitted from the nucleus with the same kinetic energy each time from a particular radioactive isotope. The nucleus can be said to have undergone the same certain change of energy for each emission, consistent with quantum theory.

Beta particles, which are fast electrons, are found to be emitted with a range of kinetic energies but momentum did not appear to be conserved. For example, the direction of an emitted beta particle and the recoil of the nucleus did not lie in opposite directions. In 1930 **Pauli** proposed that a second particle must be emitted with the beta particle which could share the energy emitted from the nucleus and also conserve the momentum. This particle is now known to be an **anti-neutrino**. **Neutrinos** and **anti-neutrinos** have **very little mass** and have **no charge**, and are not easy to detect.

The equation for beta decay is: $n \rightarrow p + e^- + \bar{\nu}$ [ν bar = nu bar]

Inside the nucleus a neutron turns into a direction of nuclear recoil proton with emission of a beta particle and its accompanying anti-neutrino.

This discovery of sub-nuclear particles raised the question about other possible particles. It also showed that if a neutron decays by emitting particles, it cannot be said to be a fundamental particle.

Sub-nuclear particles

Physicists have discovered over 200 new particles in **cosmic rays** and **particle accelerators**. They are named with letters from the Greek and Roman alphabets. Most are made from what are considered the fundamental particles, the **leptons** and **quarks**.

LEPTONS

There are six electron-like particles called **leptons**. Three are negatively charged and each of these has an associated neutral charged particle, its **neutrino**.

Generation 1   The electron (e) and its electron neutrino (ν_e)
Generation 2   The muon (μ) and its muon neutrino (ν_μ)
Generation 3   The tauon (τ) and its tauon neutrino (ν_τ)

Note the tauon is also called the tau lepton or tau particle or just the tau.

Generation 1 particles are found in ordinary matter, but generation 2 and 3 particles are mainly found in cosmic rays and particle accelerators.

The **electron** is the smallest charged particle we know of and is very stable. **Muons** have mass over 200 times greater than electrons and a lifetime of about $2\mu s$. Muons can be created in cosmic rays at different heights above the Earth. The **tauon** was discovered in high-energy particle collisions. It has a mass about 3500 times the mass of the electron and 17 times that of the muon. Its lifetime is 100 000 times shorter than that of the muon.

Neutrinos were produced in abundance in the early universe and our Sun produces millions of neutrinos. Their mass is at least a million times less than an electron.

contd

Antimatter gives a further six leptons. The **positron, anti-muon** and **anti-tauon** all have a charge of +1, and their anti-neutrinos have 0 charge.

Generation 1   The positron ($\bar{e}^+$) and its anti-electron neutrino ($\bar{v}_e$)
Generation 2   The anti-muon ($\bar{\mu}$) and its anti-muon neutrino ($\bar{v}_\mu$)
Generation 3   The anti-tauon ($\bar{\tau}$) and its anti-tauon neutrino ($\bar{v}_\tau$)

The electron's antiparticle, the positron, is identical in mass but positive, and was the first antimatter particle discovered.

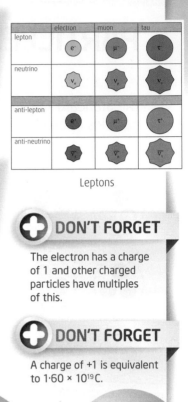

Leptons

QUARKS

In matter, there are six further particles which we consider fundamental: **quarks**. Quarks fall into the three generations and have not been found to exist on their own. **Quarks only exist in bound states.** Quarks have fractions of charges and must combine with others to form other particles with charge 1 or a multiple. They can also combine to give a charge of zero.

Generation 1   The up (u) and the down (d)
Generation 2   The charm (c) and the strange (s)
Generation 3   The top (t) and the bottom (b)

The up, charm and top all have a charge of $+\frac{2}{3}$. The down, strange and bottom all have a charge of $-\frac{1}{3}$.

Gell-Mann proposed the quarks and their three generations. He said that protons and neutrons were made of **up** and **down** quarks.

A **proton** has **two up** and **one down** quark. The charge is:
$(2 \times +\frac{2}{3}) + (1 \times -\frac{1}{3}) = +1$ positive.

A **neutron** has **two down** and **one up** quark. The charge is:
$(2 \times -\frac{1}{3}) + (1 \times +\frac{2}{3}) = 0$ neutral.

Calculate the charges on an anti-proton and an anti-neutron.

Only up and down quarks are found inside atoms of ordinary matter. The other quarks are made in particle accelerators and may also exist inside stars.

For antimatter we have three generations of quarks:

Generation 1   The anti-up (u) and the anti-down (d)
Generation 2   The anti-charm (c) and the anti-strange (s)
Generation 3   The anti-top (t) and the anti-bottom (b)

The anti-up, anti-charm and anti-top all have a charge of $-\frac{2}{3}$. The anti-down, anti-strange and anti-bottom all have a charge of $+\frac{1}{3}$.

Each successive generation of quarks have increasing mass and higher energy. The high energy quarks cannot exist in our universe for long.

Baryons and mesons

Particles that are made up of three quarks are called **baryons**. The neutron and the proton are baryons. Many other baryons can be made but their lifetime is a tiny fraction of a second. The Lambda (Λ^0) particle is made of an up, down and strange quark, but its lifetime is given as only 2.6×10^{-10} s.

Particles that are made up of two quarks are called **mesons**. Mesons are made from a quark anti-quark combination. They are unstable and short lived.

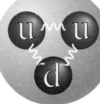

Proton   Neutron

baryons

proton   lambda
up, up,   up, down,
down   strange

mesons

pion   kaon⁰
up &   down &
anti-down   anti-strange

THINGS TO DO AND THINK ABOUT

Positron emission is similar to beta emission: $n \rightarrow p + e^+ + v$. It's used in medicine (PET scans).

Leptons, quarks and force carrier particles are the fundamental particles which form the basis of the Standard Model (used to explain our world with more precision than ever before).

There are 12 leptons and 12 quarks in total including the antiparticles.

PARTICLES AND FORCES

The Standard Model groups particles according to specific classifications, which relate how the particles interact.

HADRONS

We have discovered that many **composite particles** can be found made of quarks.

These particles are **hadrons** (or heavy particles). Hadron is from a Greek word for heavy.

Baryons are made of **three quarks**.

Mesons are made of **two quarks**.

The electric charge of a hadron is an integer number.

Baryons are made of three matter quarks or three antimatter quarks. The neutron and proton are stable and can be long-lived. Others have a very short lifetime.

- The proton is stable and long-lived whether in the nucleus or as a free particle.
- The neutron is stable in the nucleus and long-lived (~900s) when free.

Other baryons are Lambda, Sigma, Delta, Xi Cascade and Omega.

Mesons are made from a matter and antimatter pair.

- Mesons are unstable.
- Mesons have a very short lifetime.
- Pions (π) are made from matter and antimatter up and down quarks.
- J/Psi (J/ψ) particles are made from matter and antimatter charm quarks.

Other mesons are Kaon, Eta, Rho, Omega, Phi, D, B and Upsilon.

Fermions

Leptons and **quarks** along with **baryons** (made of three quarks) are **fermions**. This term means they obey Pauli's Exclusion Principle and cannot be in the same quantum state.

BOSONS

How does charge or matter attract or repel? How can a force or interaction act at a distance? We may say particles are in a field but this concept does not tell us what a field is. It is now thought that attraction or repulsion acts by exchange of particles.

The **force-mediating** or **exchange particles** are **bosons** (gluons, W- and Z-bosons, and photons). Bosons can be in the same quantum state and do not follow Pauli's exclusion principle. The forces or interchanges that physicists recognise in **order of strength** are

- the **strong (nuclear) force** with exchange particles the **gluons**. This holds the quarks together, whether inside a baryon or meson, or between these particles, and so it holds the nucleus together. As distance increases, so does the force, like an elastic band! The range is that of a neutron or 10^{-15} m.
- the **electromagnetic** interchange takes place with the massless **photon**, a particle-like wave which we study later.
- the **weak (nuclear) force** with exchange particles the **W$^\pm$ and Z** bosons, acts on quarks and leptons so acts with neutrino interactions. The range of 10^{-18} m is about 0·1% the diameter of a proton, but these bosons have much greater mass.

DON'T FORGET

Hadrons are not fundamental particles, but quarks are thought to be.

DON'T FORGET

Baryons and mesons are groups of hadrons.

DON'T FORGET

Mesons, made of two quarks, are also bosons.

contd

- **Gravitation** interchanges are thought to take place with an exchange particle called the **graviton**, but this particle has never been found. Due to this, and because of its much weaker effect compared with the other forces, it is not yet included in the Standard Model.

DON'T FORGET

Gravitational and electromagnetic ranges are infinite.

THE STANDARD MODEL

The **Standard Model** is the name given to the theory of **fundamental particles** and how they **interact**. It explains existing particles and helps predict new particles. The particles we have been studying and their exchange particles can be summarised in the table.

VIDEO LINK

Watch the clips on www. brightredbooks.net for more information on this topic.

The Higgs boson

The **fundamental scalar boson** in the Standard Model is the **Higgs boson**. This was predicted by Professor Peter Higgs of Edinburgh University in 1964 but not confirmed till March 2013 by the Large Hadron Collider, earning Higgs the 2013 Nobel prize. The Higgs boson helps to explain why the gauge bosons of the weak field have mass but the photon does not.

THE STANDARD MODEL				
Fundamental particles	Fermions	Quarks	u, d, c, s, t, b, $\bar{u}$, $\bar{d}$, $\bar{c}$, $\bar{s}$, $\bar{t}$, $\bar{b}$	
		Leptons	e-, e+, μ, $\bar{μ}$, τ, $\bar{τ}$, v_e, $\bar{v}$, $v_μ$, $\bar{v}_μ$, $v_τ$, $\bar{v}_τ$	
	Bosons	Gauge	γ, g, W$^{\pm}$, Z	
		Scalar	$\bar{H}^0$	
	Others		To be continued.....	
Composite particles	Hadrons	Baryons	n, p, Δ, Λ, Σ, Ξ, Y	=>Fermions
		Mesons	π, ρ, η, ή, φ, ω, J/ψ, Υ, θ, K, B, D, T	=>Bosons
	Others		To be continued.....	

Particle exchange diagrams

Particle exchange diagrams show how particles interact by force exchange particles. The use of these diagrams can help our understanding of the particles and forces.

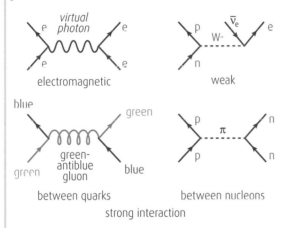

electromagnetic   weak

strong interaction
between quarks   between nucleons

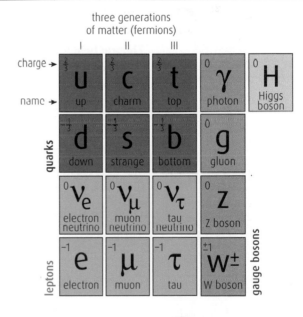

three generations of matter (fermions)

Beta decay updated

How does a neutron change into a proton with release of the beta particle and its anti-neutrino? We can see a down quark has become an up quark changing the neutron to a proton. W- is a force-carrier particle. The electron takes away the charge. The electron and anti-neutrino are created as a pair. Our knowledge is shown in the new diagram.

ONLINE

Head to www. brightredbooks.net and follow the links to read more about fundamental forces and repulsive exchange.

THINGS TO DO AND THINK ABOUT

The Standard Model and particle physics still has many questions which you can follow online.

ONLINE TEST

Take the test on particles and forces at www. brightredbooks.net

ELECTRIC FIELDS

Beside exchange particle theory we learn how concepts of fields are used to explain the forces on charged particles. We learn examples of electric field patterns and how charges behave in electric fields.

ELECTRIC FIELDS

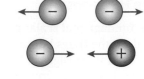

There are two types of charge: **positive** and **negative**.

Like charges repel. Unlike charges attract.

The region round an electric charge is an **electric field**.

In an **electric field**, an **electric charge** experiences a **force**.

The **electric field strength** is a measure of the **force** on a **unit charge**.

It sometimes helps to compare **electric fields** with **gravitational fields**:

In a **gravitational field**, a **mass** experiences a **force**.

The **gravitational field strength** is a measure of the **force** on a **unit mass**.

Gravitational field lines show the **direction** that a **mass** will move in the **gravitational field** due to the force on it.

Electric field lines show the direction that a **free positive charge** will move in an **electric field**. The **direction of a field** round a charge depends on whether its sign is **positive** or **negative**.

The **electric field** causes the **free charges** in it to **move**.

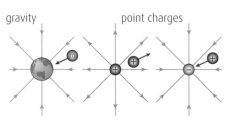
gravity   point charges

Radial fields

The field strength in a radial field decreases with distance.

Uniform fields

The field strength in a uniform field is constant.

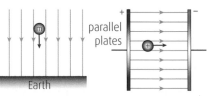
parallel plates
Earth

Combining fields

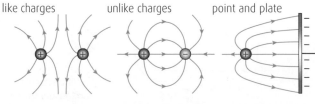
like charges   unlike charges   point and plate

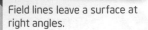
WORK DONE

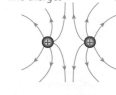

When **work is done** to push a charge against a field, the charge has a **gain** in **potential energy**.

When the field **does work** on the charge, the charge **accelerates** and gains **kinetic energy**.

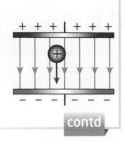

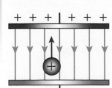

contd

Work (*W*) is done when a **charge (*Q*) is moved** in an **electric field**.

The amount of work done depends on

- the **size of the charge** being moved, *Q*
- the **size of the voltage *V*** creating the electric field.

$W = QV$ or $E_w = QV$

A charge accelerated parallel to a field will gain kinetic energy from the work done by the field.

$QV = \frac{1}{2} mv^2$

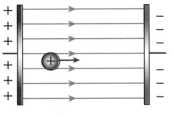

Example:

An electron is accelerated by a voltage of 2000 V. What is its final speed?

Solution:

$$QV = \frac{1}{2}mv^2 \qquad v = \sqrt{\frac{2QV}{m}} \qquad v = \sqrt{\frac{2 \times 1 \cdot 6 \times 10^{-19} \times 2000}{9 \cdot 11 \times 10^{-31}}} \qquad v = 2 \cdot 65 \times 10^7 \, ms^{-1}$$

Potential difference

$V = \frac{W}{Q}$

The **potential difference (p.d.)** measured in volts (V) between two points is a measure of the **work done (*W*)** in moving **1 coulomb of charge (*Q*)** between the two points.

If **1 joule of work** is done moving **1 coulomb** of **charge** between two points, the **potential difference** between the two points is **1 volt**.

For example

- $3V = 3 \, JC^{-1}$, 3 J of work is done for each 1 C of charge
- $900V = 900 \, JC^{-1}$, so 900 J of work is done for each 1 C of charge.

ELECTRIC FIELDS ACROSS A CONDUCTOR

An **electric field** applied to a **conductor** causes the free electric **charges** in it to **move**.

A 12 V battery supplies 12 J of work to each 1 C of charge. These equations also apply to electric circuits:

$W = QV$ $\quad V = \frac{W}{Q}$

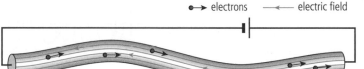

● → electrons   ← electric field

THINGS TO DO AND THINK ABOUT

1   Useful data: check the data page in the exam!

PARTICLE	SIGN	CHARGE, Q (C)	MASS, m (kg)
electron	negative −	$1 \cdot 60 \times 10^{-19}$	$9 \cdot 11 \times 10^{-31}$
proton	positive +	$1 \cdot 60 \times 10^{-19}$	$1 \cdot 673 \times 10^{-27}$

2   If 6000 J of work are required to move 3 C of charge through an electric field, what is the potential difference across the field?

$V = \frac{W}{Q} = \frac{6000}{3} = 2000 \, V$

3   Electric field strength gives the force that would be exerted on a test unit charge, even if the test charge is not present.

DON'T FORGET

Work done = energy transferred.

DON'T FORGET

The field does work, $W = QV$. The charge gains energy, $E_k = \frac{1}{2}mv^2$. These equations will not be put together for you but you should understand how one converts to the other.

DON'T FORGET

Work done on unit charge gives a definition of the volt.

ONLINE

Head to www. brightredbooks.net to learn more about electric fields and non-contact forces.

VIDEO LINK

Watch the clips at www. brightredbooks.net for more on electric fields.

ONLINE TEST

Take the test on electric fields at www. brightredbooks.net

MAGNETIC FIELDS

The concepts of fields are used to explain the forces on magnetic poles and moving charge. We learn examples of magnetic field patterns and how moving charges behave travelling through magnetic fields.

PERMANENT FIELD PATTERNS

Magnets have two poles, north and south. Individual poles cannot be isolated.

The north pole of a magnet seeks the Earth's magnetic pole near geographic North. Confusingly, geographic North is actually a magnetic south pole.

Like poles repel. Unlike poles attract.

The region round magnetic poles is a **magnetic field**.

In a **magnetic field**, poles experience a force. A **north pole** experiences a **force** in the direction shown by the field lines, a south pole in the opposite direction. A compass lines up in the direction of the field lines and can be used for navigation.

Our Earth's magnetic field helps protect us from charged particles heading towards Earth.

The field round a bar magnet is similar to the Earth's magnetic field.

The field is stronger when the lines are closer.

The field is weaker when the lines are further apart.

The field will cancel between two like poles. The magnetic field strength is zero. A north and/or south pole placed between these two like poles will not move.

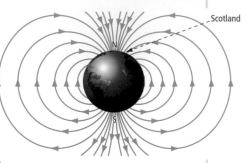
Scotland

> **DON'T FORGET**
>
> The magnetic field exists in three dimensions.

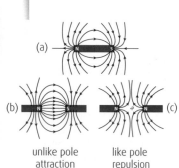

(a)

(b)   (c)

unlike pole attraction   like pole repulsion

Permanent magnetic field around a bar magnet.

OERSTED'S DISCOVERY

In 1820 Hans Christian Oersted discovered that a magnetic field exists around a wire carrying a current. The **strength** of the magnetic field

- **increases** with the size of the **current**
- **decreases** as the **distance** from the wire **increases**.

The magnetic field lines form **concentric circles** around the wire. You can think of a continuous but weakening field as you move away from the wire.

The field lines will be **anti-clockwise** when **electron flow** is **away** down the wire. A memory aid is **bye-bye anti**, (wave goodbye to aunty is an anti-clockwise field when electrons move away). Alternatively, you can use the **left hand grip rule** instead.

The field lines will be **clockwise** when electron flow is **towards us** in a wire.

> **DON'T FORGET**
>
> Conventional current, considered a flow of positive charge, reverses the rules. Current up gives an anti-clockwise field and we would use the **right hand grip rule** instead.

CHARGE IN A MAGNETIC FIELD

All electric charges have an electric field. A **moving charge** produces a **magnetic field** in addition to the electric field. A current generates a magnetic field as it is a flow of charges. Moving charge generates its magnetic field whether it is in a conductor or moving through space.

contd

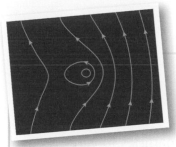

A current-carrying wire crossing a magnetic field will experience a force due to the interaction of the magnetic field around the wire and the field it is crossing. The fields are in the same direction on one side of the wire and add, but are in the opposite direction on the other side so cancel. The wire moves sideways.

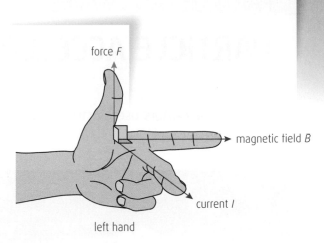

force *F*

magnetic field *B*

current *I*

left hand

Left hand rule for positive charge

Consider a single charge *q*, moving with velocity *v*, entering a magnetic field *B* perpendicular (at right angles) to that field.

Charge experiences a force whose direction is **perpendicular** to both velocity *v* and magnetic field *B*.

Use **Fleming's left hand rule** for **positive** charges moving across a magnetic field.

First finger - **F**ield

Se**C**ond finger - **C**urrent

Thu**M**b - **T**hrust or **M**otion

Use the same rule but with your **right** hand for **negative** charges moving across a magnetic field.

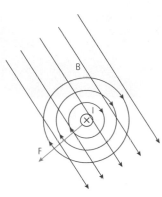

thrust (force)

magnetic field

current

Right hand rule for negative charge

The strength of the force increases with

- the magnetic field strength
- the size of the charge
- the velocity of the particle.

B

F

Positive charge at velocity *v* into the page.

Apply the left and right hand rules to these two charged particles. Find the direction of the force on each particle. (The crosses indicate the direction of the magnetic field is into the page.)

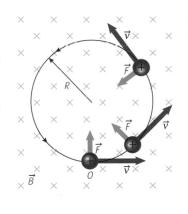

If a charge stays in a magnetic field long enough it may end up trapped in a **circular path**. The radius of the path will also depend on the **mass** of the particle.

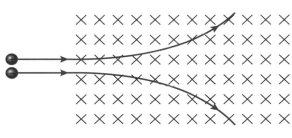

DON'T FORGET

The direction changes, the velocity does not.

ONLINE

Revise this topic further by following the link at www.brightredbooks.net

VIDEO LINK

Check out the clip at www.brightredbooks.net to see the alternative RH rule.

ONLINE TEST

Head to www.brightredbooks.net and take the test on magnetic fields.

THINGS TO DO AND THINK ABOUT

Motors and generators are opposites of each other. Find out how these devices use charges in magnetic fields to produce their effects.

Electric fields and magnetic fields used to be considered independent branches of physics. Now these branches are united into electromagnetism.

PARTICLE ACCELERATORS

Particle accelerators use electric or magnetic fields to accelerate charged particles to very high speeds.

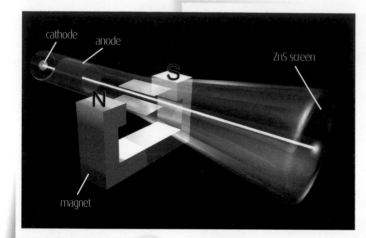

cathode
anode
ZnS screen
S
N
magnet

ELECTROSTATIC ACCELERATORS

In the laboratory, **charged particles** can be accelerated in a vacuum tube. A low voltage (e.g. 6 V) applied to a heater or cathode produces a cloud of **electrons** which are then accelerated through an **electric field** between the **cathode** and an **anode**. Further vertical and horizontal **electric fields** can be created to deflect the beam of electrons up and down or from side to side.

A high voltage of a few thousand volts is used to create suitable electric fields. In a short distance the electrons may reach speeds in the order of magnitude of $10^7 \, ms^{-1}$. When electrons collide with a phosphor screen it fluoresces.

Magnetic fields may be used as an alternative to electric fields. These are created by two Helmholtz coils placed on either side of the vacuum tube. When a current is in the coils a magnetic field **crosses** the path of the electrons.

Some experiments use both electric and magnetic fields to investigate the charge and mass of electrons.

Electron beams were originally known as **cathode rays** before it was discovered that these rays were actually negatively-charged electrons.

LINEAR ACCELERATORS

Cathode ray tubes are one type of **linear accelerator**. The largest linear accelerator built is in California and is 3·2 km long. With larger accelerators, particle physicists measure the energy reached as an alternative to the velocity. The energy of one electron accelerated through a potential difference of 1 volt is: $W = qV = (1\cdot60 \times 10^{-19}) \times 1 = 1\cdot60 \times 10^{-19} \, J$.

1 electron volt, 1 eV = 1·60 × 10⁻¹⁹ J

Linear particle accelerators, or **Linacs**, are able to accelerate heavy ions to high energies without the need for very strong magnetic fields to keep these ions in a fine beam.

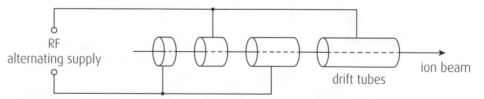

RF alternating supply
drift tubes
ion beam

Repeated acceleration of charged particles by **electric fields** achieves very high energies. Linear accelerators are made with a series of metal tubes of **increasing length** to accommodate the **increasing velocity** of the particles. The **time** spent in each tube remains the **same**. The charged particles are **accelerated** in the **electric fields** between the tubes, not in the tubes. An **a.c. supply** is used to **alternate the polarity** of the tubes as charged particles accelerate between each pair. Thus when a group of electrons is between any two tubes, the tube behind will be negative and the tube in front will be positive.

CYCLOTRONS

In a **cyclotron**, **charged particles accelerate** outwards from the centre along a **spiral path**. Cyclotrons have increased in size since being invented in 1932 and the largest built, at 18 m diameter, does not require the length of the Linacs.

A **high frequency a.c. voltage** is applied across the dees. The charged particle is **accelerated in the gap** between the dees by an **electric field**.

A **magnetic field** is applied **perpendicular** to the dees which bends the particles into a **circular path**. As the **velocity increases** the radius of the circular path increases.

At high energies relativistic effects are taken into account.

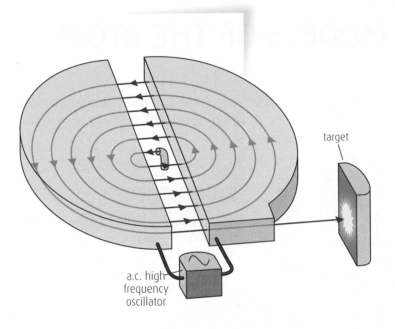

target

a.c. high frequency oscillator

Cyclotron

SYNCHROTRON

In a **synchrotron**, **charged particles accelerate repeatedly** around a **ring of constant radius**. As the **velocity increases**, the **energy increases**. At high relativistic speeds approaching the speed of light, the extra energy shows as a large **increase in mass**. The **frequency** of the **supply** which increases the energy, and the **frequency** of the **magnetic field** which maintains the particles in their curved path, are both **altered** as the energy increases.

The **Large Hadron Collider**, at CERN on the Swiss–French border, has the world's largest **synchrotron** with a 27 km circumference. **Two rings** with particles accelerating in **opposite directions** cross at several points so that particle collisions take place. Huge **detectors** trace the path of the **created** sub-nuclear particles.

CERN synchrotron is under the Swiss–French border

Nuclear particles such as protons as well as **heavier ions** in each ring can be accelerated to 7 TeV so the resultant collision from each beam takes place at 14 TeV. The largest collision before CERN was at 1 TeV in the **Tevatron** accelerator in the USA.

The accelerator at CERN is actually a combination of linear and synchrotron accelerators and the particles travel at different depths as they make their way in stages to the highest energies.

ONLINE

Read more about electron deflection and Thomson's experiment at www.brightredbooks.net

VIDEO LINK

Learn more about CERN by watching the clips at www.brightredbooks.net

THINGS TO DO AND THINK ABOUT

On 4 July 2012, the ATLAS and CMS experiments at CERN's Large Hadron Collider announced they had each observed a new particle around 126 GeV consistent with the Higgs boson as predicted by the Standard Model.

As well as particle research, accelerators are used in industrial processing and radiotherapy in medicine.

ONLINE TEST

Want to test your knowledge of particle accelerators? Head to www.brightredbooks.net

MODELS OF THE ATOM

Scientists and philosophers have been considering models of the atom for over 2000 years. Understanding the development of the different models will help you put the Higher course requirements in context.

ATOMIC THEORY

The questioning of the nature of matter has been ingrained in philosophical thinking for thousands of years. A belief that repeated dividing of matter into smaller and smaller parts would eventually lead to small, **indivisible** elements was recorded by the Greek philosopher **Democritus**, born around 460 BC. Democritus referred to his basic matter particles as '**atomos**', meaning indivisible.

Experimental evidence convinced **John Dalton**, a physicist and chemist, that elements combine in fixed ratios, which correspond to elements having building blocks of fixed mass. He developed his **atomic theory** in 1803.

- Elements consist of small particles called **atoms**.
- An **element** has only **one** type of atom.
- Different atoms have different **atomic weights**.
- Atoms combine to form **compounds**.
- Atoms are not created or destroyed in reactions.

SUB-ATOMIC PARTICLES

J.J. Thomson was studying cathode rays inside vacuum tubes when he proposed they were made of particles from inside atoms. In 1897 on discovering 'corpuscles' (electrons), he said

'I can see no escape from the conclusion that cathode rays are charges of negative electricity carried by particles of matter'.

He calculated their charge-to-mass ratio and found that electrons were much smaller than the size of an atom. Thomson's discovery of the **electron** was the first indication that atoms contained **sub-atomic particles**.

In 1910, Thomson proposed the **plum-pudding model**, with thousands of tiny negatively charged corpuscles inside a cloud of massless positive charge.

negative electron 'plums'

positive 'pudding'

Experimental evidence improves our models and in 1911 Thomson's former research student, **Ernest Rutherford**, came up with an improved theory of the atom, after having supervised two young colleagues (Geiger and Marsden) performing an experiment.

Ernest Rutherford was the son of a Scottish wheelwright, was born in New Zealand and succeeded J.J. Thomson as Professor of Physics at Cambridge. Many other famous scientists owe their discoveries to his guidance and you could learn more about the history of atomic physics by looking at Rutherford's career.

Radioactivity was a recent discovery and Rutherford was contributing to the study of radioactive particles. Rutherford suggested that alpha particles could be used to probe atoms. He expected alpha particles would have some slight scattering from the charges in the Thomson model of the atom.

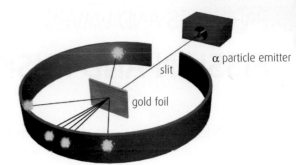

α particle emitter

slit

gold foil

detecting screen

GEIGER AND MARSDEN'S EXPERIMENT

The Geiger and Marsden experiment is one of the most important experiments in physics. A stream of **alpha particles** (charge +2, mass 4 amu) was fired at a **thin sheet** of **gold foil** in a vacuum.

Geiger and Marsden observed the scattering of the alpha particles using a fluorescent screen.

Based on Thomson's plum-pudding model, many scattering angles were expected, but a different result was found:

* **most** particles **passed straight through**
* only a **few** were **deflected** through large angles
* about 1 in 8000 **came backwards**.

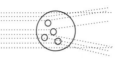

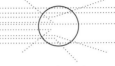

expected result   actual result

NEW ATOMIC MODELS

When Geiger and Marsden reported what they had observed, **Rutherford** suggested the explanations:

* most particles pass through the foil (>100 atoms thick), **so most of the atom must be empty space**

* a few were deflected or bounced back, **so most of the mass and positive charge of an atom is concentrated in a very small volume**.

gold atom   alpha particles

Rutherford described it as the most incredible event of his life 'as if you fired a 15-inch shell at a piece of tissue paper and it came back and hit you'.

In 1911, Rutherford proposed his planetary model or **nuclear model**. This model suggested that the **atom** might resemble a tiny solar system, with a **massive, positively charged centre** or **nucleus** circled by only a **few electrons**.

* The nucleus has a relatively small diameter compared with that of an atom.
* Most of the mass of an atom is concentrated in the nucleus.

In 1913, **Niels Bohr** proposed his **orbit model**. Based on the work you will study with emission lines, photons and energy levels, Bohr confined **electrons** to orbit **in certain energy levels** or **shells**. The electrons **radiate energy** only in **transitions**.

In 1924–26, Louis de Broglie and Erwin Schrödinger proposed the **electron cloud** or **quantum mechanical model**, in which electrons exist in wavelike orbitals. (This is beyond the Higher Physics course.)

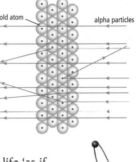

1e-

3p+

2e-

nucleus

DON'T FORGET

Based on the Standard Model our knowledge of the nucleus continues to expand.

ONLINE

Follow the link at www.brightredbooks.net to learn more about models of the atom.

VIDEO LINK

Head to www.brightredbooks.net to watch clips on this topic.

ONLINE TEST

Head to www.brightredbooks.net and test yourself on models of the atom.

THINGS TO DO AND THINK ABOUT

According to the quantum theory developed by Albert Einstein and others, electrons act like particles under some conditions and as waves under other conditions. The wave character of electrons was experimentally indicated by J.J. Thomson's own son, G.P. Thomson.

Rutherford discovered the proton in 1918 and Chadwick the neutron in 1932.

FISSION AND FUSION

Here we learn to use nuclear equations for radioactive decay and describe fission and fusion reactions.

PARTICLE	MASS (AMU)	CHARGE	SYMBOL
proton	1	+1	p
neutron	1	0	n
electron	$\frac{1}{1840}$	−1	e

DECAY OF RADIONUCLIDES

The structure of the atom is based on protons, neutrons and electrons and some other particles from our Standard Model.

A **radionuclide** is an isotope that decays radioactively.

Alpha, **beta** and **gamma radiations** all **emit** from the **nucleus** of an atom.

RADIATION	NATURE	MASS (AMU)	SPEED	CHARGE	SYMBOL	DEFLECTION IN A MAGNETIC OR ELECTRIC FIELD
alpha	He nucleus	4	slow	+2	α	yes
beta	fast electron	$\frac{1}{1840}$	0.9 c	−1	β	yes
gamma	e-m wave	0	c	0	γ	no

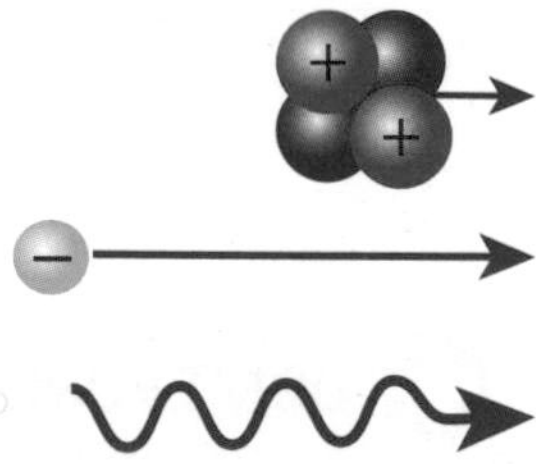

Alpha decay: α

The daughter is a different element. $^{A}_{Z}X \rightarrow \, ^{A-4}_{Z-2}Y + \, ^{4}_{2}\alpha$

Beta decay: β

A neutron decays into a proton, an electron (beta) and an anti-neutrino. $^{1}_{0}n \rightarrow \, ^{1}_{1}p + \, ^{0}_{-1}\beta$

Gamma decay: γ

No change in isotope, energy loss by e-m radiation.

The **alpha** particle is identical to a helium nucleus but it is not helium, as it has no orbiting electrons.

The **beta** particle is a fast electron, which also comes from the nucleus.

A **gamma** decay is a burst of energy from the nucleus.

DON'T FORGET

mass number, A = number of protons + neutrons

$^{56}_{26}Fe$

atomic number, Z = number of protons

DON'T FORGET

Isotopes of an element have the same atomic number but different mass number.

DON'T FORGET

Fission = splitting
fusion = joining

DON'T FORGET

Fission and fusion are both **nuclear** processes.

NUCLEAR EQUATIONS

Nuclear equations can be written for any nuclear reaction or decay.

The **total** of the **mass numbers** and the **atomic numbers** must be the same on both sides of the equation.

Examples:

Practice balancing these equations before checking the answers that follow. You will need a periodic table to find the missing element.

a $^{42}_{19}K \rightarrow \, ^{0}_{-1}e + ?$

b $^{226}_{88}Ra \rightarrow \, ^{4}_{2}He + ?$

c $^{6}_{3}Li + \, ^{1}_{0}n \rightarrow \, ^{0}_{-1}e + \, ^{4}_{2}He + ?$

d $^{239}_{94}Pu \rightarrow \, ^{4}_{2}He + ?$

e $^{1}_{1}H + \, ^{3}_{1}H \rightarrow ?$

f $^{235}_{92}U + \, ^{1}_{0}n \rightarrow ? \, ^{1}_{0}n + \, ^{139}_{53}I + \, ^{95}_{39}Y$

g $^{14}_{6}C \rightarrow \, ^{14}_{7}? + \, ^{0}_{-1}$

contd

ONLINE

Read more about the difference between fission and fusion at www.brightredbooks.net

Solutions:

a $^{42}_{20}Ca$ **b** $^{222}_{86}Rn$ **c** $^{3}_{2}He$ **d** $^{235}_{92}U$ **e** $^{4}_{2}He$ **f** 2 **g** Nitrogen

Positron emission

Here is an equation for positron emission: $^{11}_{6}C \rightarrow {}^{11}_{5}B + {}^{0}_{1}e^{+} + {}^{0}_{0}\nu_{e}$

We see how the emission of the neutral neutrino does not affect the balancing of the equation, just makes the equation more complete.

VIDEO LINK

Check out the clips at www.brightredbooks.net

NUCLEAR FISSION

Fission is the **splitting** of a large **nucleus**.

In **fission**, a **nucleus** with a **large mass number splits** into **two nuclei** of **smaller mass numbers**, usually with the release of **neutrons**. Energy is **released**.

Spontaneous fission

Fission may be **spontaneous**, with each fission occurring at **random**. The **half-life** will be **constant** for a large number of atoms.

$^{256}_{100}Fm \rightarrow {}^{140}_{54}Xe + {}^{112}_{46}Pd + {}^{4}_{0}{}^{1}n + energy$

Induced fission

Fission may be **induced** by **neutron** bombardment. An **incident neutron** can **stimulate** the **fission** of a **nucleus** with a **large mass number**.

In the following reaction, the U^{235} momentarily becomes U^{236}, but this is unstable and immediately undergoes fission.

$^{1}_{0}n + {}^{235}_{92}U \rightarrow {}^{141}_{56}Ba + {}^{92}_{36}Kr + {}^{3}_{0}{}^{1}n + energy$

Induced fission is used in the **reactors** in **nuclear power stations**.

NUCLEAR FUSION

Fusion is the **joining** of **nuclei**.

In **fusion**, two nuclei **combine** to form a **nucleus** of **larger mass number**. The nuclei that fuse together are usually very **small**.

A large amount of **energy** is released, and **no radioactive waste** is produced in the reaction.

There is a **virtually unlimited** amount of the isotopes of hydrogen needed for **fusion** in seawater, and no greenhouse gases are emitted.

$^{2}_{1}H + {}^{3}_{1}H \rightarrow {}^{4}_{2}He + {}^{1}_{0}n + energy$

Fusion is the energy source of the **Sun** and the **stars**, but physicists are still working on the design of **nuclear-fusion reactors** for Earth.

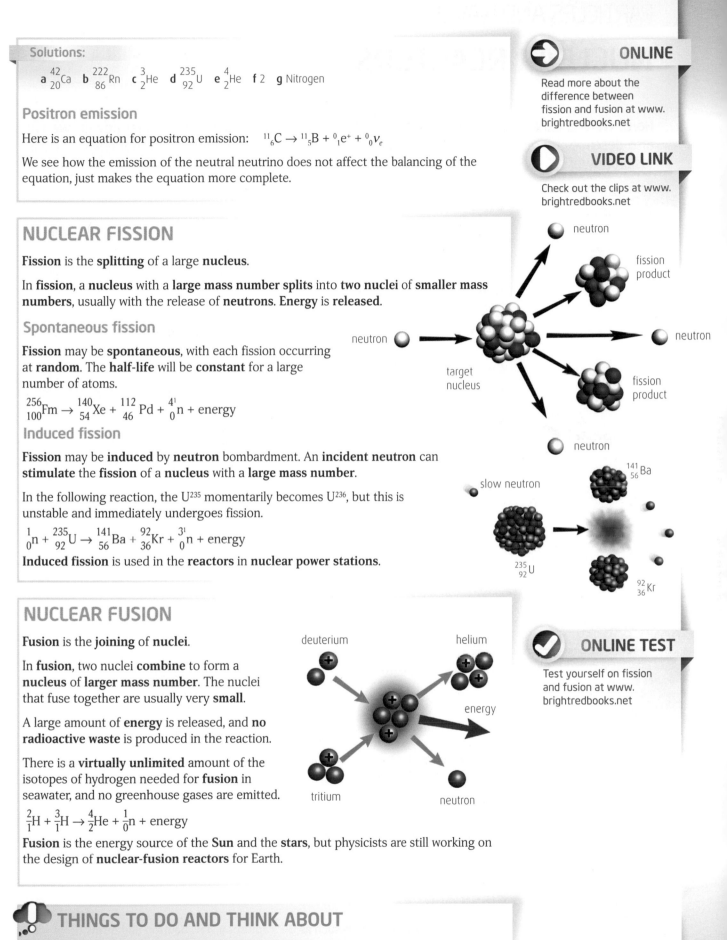

ONLINE TEST

Test yourself on fission and fusion at www.brightredbooks.net

THINGS TO DO AND THINK ABOUT

Describe the **difference** between fission and fusion.
Describe both fission and fusion in terms of **nuclei** and **mass numbers**.

NUCLEAR REACTORS

Here, we will explain the relevance of Einstein's most famous equation to nuclear power. We will also describe the operation of nuclear fission reactors and some problems with obtaining nuclear fusion power.

EINSTEIN AND ENERGY

During **fission** or **fusion**, the number of **nucleons** (particles of the nuclei) is the **same** before and after the reaction. The total **mass number** is conserved.

The **total mass** of the individual particles before and after the reaction is **not the same**. The **mass of the products** is always **less** than the **mass of the starting materials** or **reactants**. The missing mass is called the **lost mass**.

Einstein proposed that the **mass** and **energy** are **equivalent** and that the lost mass is turned into released energy, using his famous relationship:

$$E = mc^2$$

The products of **fission** and **fusion** acquire large amounts of **kinetic energy**.

Calculating the energy released by fission or fusion

- Find the difference in mass between each side of the equation.
- Calculate the energy released from the lost mass using Einstein's equation.

> **DON'T FORGET**
>
> The mass of a nucleus is always less than the total mass of its individual nucleons.

Example:

How much energy is released from a single nucleus in the following fission reaction?

$$^{1}_{0}n + ^{235}_{92}U \rightarrow ^{138}_{55}Cs + ^{96}_{37}Rb + 2^{1}_{0}n$$

Solution:

Mass before:	$^{1}_{0}n$	1.6750×10^{-27} kg
	$^{235}_{92}U$	3.9014×10^{-25} kg Total $= 3.91815 \times 10^{-25}$ kg
Mass after:	$^{138}_{55}Cs$	2.2895×10^{-25} kg
	$^{96}_{37}Rb$	1.5925×10^{-25} kg
	$^{1}_{0}n$	1.6750×10^{-27} kg
	$^{1}_{0}n$	1.6750×10^{-27} kg Total $= 3.9155 \times 10^{-25}$ kg

Lost mass $= 2.65 \times 10^{-28}$ kg
Energy released: $E = mc^2 = 2.65 \times 10^{-28}(3.0 \times 10^8)^2 = 2.38 \times 10^{-11}$ J

> **DON'T FORGET**
>
> Take care not to drop figures in lost-mass calculations – they are all significant!

Example:

There are $\dfrac{1}{3.9014 \times 10^{-25}} = 2.56 \times 10^{24}$ atoms in 1 kg of uranium.

The energy in 1 kg is therefore $(2.835 \times 10^{-11})(2.56 \times 10^{24}) = 6.11 \times 10^{13}$ J
1 kg of coal releases approximately 3×10^7 J. That is about 2 million times less than 1 kg of uranium.

> **DON'T FORGET**
>
> Nuclear-fusion calculations are done in the same way.

FISSION REACTORS

When a large unstable nucleus such as U^{235} captures a slow neutron and forms U^{236} it will immediately undergo fission into two smaller nuclei along with the release of three more fast neutrons. In order to use the fast neutrons for further fission a reactor has to be designed to slow or moderate these neutrons.

contd

The main parts of a nuclear fission reactor are
- fuel rods
- moderator
- control rods.
- coolant
- containment vessel

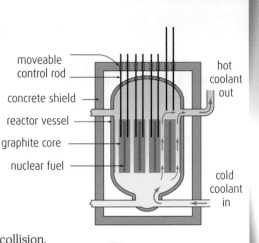

Labels: moveable control rod, concrete shield, reactor vessel, graphite core, nuclear fuel, hot coolant out, cold coolant in

Fuel rods

Pellets of enriched uranium are stacked in a rod, the mass is below critical mass, and the rods are grouped to form elements.

Moderator

A moderator such as graphite, water or heavy water surrounds the uranium elements so neutrons passing between elements can be slowed by collision.

Control rods

Control rods made of boron can be lowered or raised between the fuel rods to absorb neutrons. They are lowered to reduce the chain reaction or raised to meet demand.

Coolant

The heat from the nuclear reactions is removed using a coolant, a pressurised liquid or gas such as water or carbon dioxide. The energy is passed through a heat exchanger to a turbine and electrical generator.

Containment vessel

The reactor is surrounded by a large steel lining and concrete to absorb radiation in an accident or in the event of natural disaster.

DON'T FORGET

You should concentrate on the coolant and containment issues in nuclear fusion reactors for your exam.

ONLINE

Learn more about fission reactors at www.brightredbooks.net

FUSION REACTORS

Fusing **light nuclei** together releases huge amounts of **energy**. On Earth we have plenty of **deuterium** in sea water and **tritium** can be made from lithium. These two **isotopes of hydrogen** have only one proton and two and three neutrons. Fusion reactions are hard to achieve because of the **repulsive** nature of the **positive** charge. Temperatures of 10^8 K are required to simulate fusion which takes place in our Sun. At high temperature molecules lose their electrons and become positively charged ions, to form a **plasma**. When the nuclei are close enough the **strong nuclear force** can overcome the **electrical force**.

The **plasma** must be
- **heated** by inducing high electric currents within it
- **contained** to keep it away from the walls which would melt
- **confined** plasma density for high energy output.

Tokamak is Russian for toroidal magnetic chamber.

The neutral neutrons escape with most of the energy. This energy is absorbed to produce the heat required for use in a power station.

The ITER or International Thermonuclear Experimental Reactor is the largest being built and is to switch on in 2019.

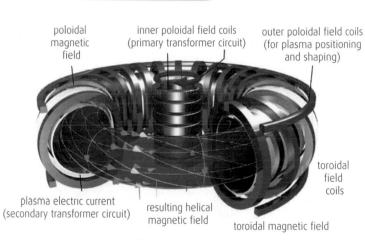

Labels: poloidal magnetic field, inner poloidal field coils (primary transformer circuit), outer poloidal field coils (for plasma positioning and shaping), toroidal field coils, plasma electric current (secondary transformer circuit), resulting helical magnetic field, toroidal magnetic field

Tokamak's circular magnetic bottle

$$^{2}_{1}\text{H} + {}^{3}_{1}\text{H} \rightarrow {}^{4}_{2}\text{He} + {}^{1}_{0}\text{n}$$

VIDEO LINK

Head to www.brightredbooks.net and check out the videos for more on this topic.

ONLINE TEST

Head to www.brightredbooks.net and test yourself on nuclear reactors.

THINGS TO DO AND THINK ABOUT

Investigate different designs of reactors, both fission and research fusion reactors.

WAVE PROPERTIES

It is important that we learn, understand and distinguish between the different properties that waves can have.

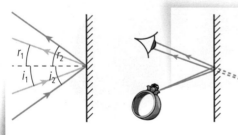

WAVES

Waves can be created by **vibrating sources**.

A wave is a **movement** of **energy**. A wave can be considered as a means of transferring energy from one point to another.

Wavelength λ is the distance in which a wave repeats.

Amplitude A is the height to a crest or trough from rest.

The **energy** (E_P and E_K) of a wave depends on its **amplitude**.

The **period T** is the time for one wave. $\quad T = \frac{1}{f}$

The **frequency f** is the number of waves that pass a point in unit time.

$$\text{frequency} = \frac{\text{number}}{\text{time}} \text{ and } f = \frac{1}{T}$$

The **frequency** of a wave is the same as the frequency of the source producing it.

The **wave speed v** is the distance travelled in unit time. $\quad v = \frac{s}{t}$

The **wave equation** relates wave speed, frequency and wavelength. $\quad v = f\lambda$

All different types of waves can show properties of **reflection**, **diffraction**, **refraction** and **interference**.

REFLECTION

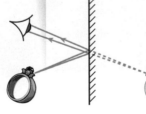

On reflection, the angle of incidence is the same as the angle of reflection. $\quad i = r$

v, f and λ are unchanged. Although the light rays **reflect**, they appear to come from behind the mirror.

DIFFRACTION

Bending called **diffraction** occurs at **edges** or **gaps**.

Circular wavefronts are produced if the gap is $\leqslant 1\lambda$.

Edges

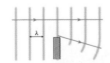

Gaps

Low-frequency long wavelengths **diffract most**

High-frequency short wavelengths **diffract least**

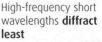

Low-frequency long wavelengths **diffract most**

High-frequency short wavelengths **diffract least**

REFRACTION

Waves **change direction** when passing from one medium to another because the waves **change velocity**.

v changes, λ changes, frequency f **stays the same**.

$$f = \frac{v_1}{\lambda_1} = \frac{v_2}{\lambda_2}$$

If the **frequency** of a light ray changed, the **colour** would change.

This ray could be travelling in either direction. Observe that the angle from the normal is smaller in the more dense material.

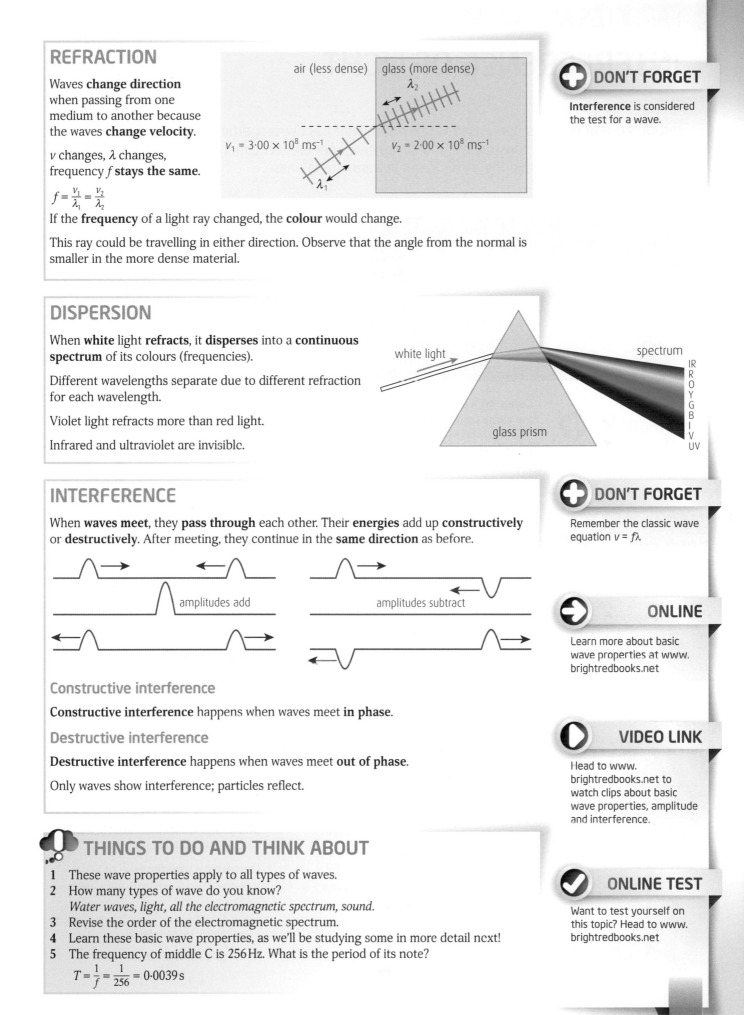

air (less dense)   glass (more dense)

λ_2

$v_1 = 3 \cdot 00 \times 10^8 \text{ ms}^{-1}$ $v_2 = 2 \cdot 00 \times 10^8 \text{ ms}^{-1}$

λ_1

DON'T FORGET

Interference is considered the test for a wave.

DISPERSION

When **white** light **refracts**, it **disperses** into a **continuous spectrum** of its colours (frequencies).

Different wavelengths separate due to different refraction for each wavelength.

Violet light refracts more than red light.

Infrared and ultraviolet are invisible.

white light

spectrum

IR R O Y G B I V UV

glass prism

INTERFERENCE

When **waves meet**, they **pass through** each other. Their **energies** add up **constructively** or **destructively**. After meeting, they continue in the **same direction** as before.

amplitudes add   amplitudes subtract

Constructive interference

Constructive interference happens when waves meet **in phase**.

Destructive interference

Destructive interference happens when waves meet **out of phase**.

Only waves show interference; particles reflect.

DON'T FORGET

Remember the classic wave equation $v = f\lambda$

ONLINE

Learn more about basic wave properties at www. brightredbooks.net

VIDEO LINK

Head to www. brightredbooks.net to watch clips about basic wave properties, amplitude and interference.

ONLINE TEST

Want to test yourself on this topic? Head to www. brightredbooks.net

THINGS TO DO AND THINK ABOUT

1. These wave properties apply to all types of waves.
2. How many types of wave do you know?
   Water waves, light, all the electromagnetic spectrum, sound.
3. Revise the order of the electromagnetic spectrum.
4. Learn these basic wave properties, as we'll be studying some in more detail next!
5. The frequency of middle C is 256 Hz. What is the period of its note?

$$T = \frac{1}{f} = \frac{1}{256} = 0 \cdot 0039 \text{ s}$$

INTERFERENCE OF LIGHT

Interference of light has been used to show that light has the properties of waves. The wave nature of light leads to constructive and destructive interference. Coherence is a special property of light waves which have the same frequency and are in phase.

YOUNG'S DOUBLE-SLIT EXPERIMENT

Thomas Young's famous **double-slit interference experiment** from 1801 showed that light had wave properties. He concluded that **light energy** travelled as a **wave motion**. Using a **single light source** with the **double slits** he saw an **interference pattern** of a **series of light and dark fringes**. The single light source produces **coherent waves**. **Diffraction** at each slit produces **circular wavefronts**, which spread out so that they meet and produce **interference**.

Coherent waves have the **same frequency**, with a **constant phase relationship** (either they are in phase or have a constant phase difference).

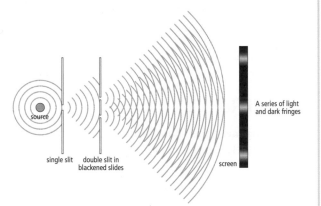

A series of light and dark fringes

single slit   double slit in   screen
              blackened slides

source

PATH DIFFERENCE

Constructive interference takes place when the two light waves are **in phase**. The waves are **in phase** when the **path difference, PD,** is an **integer** number of **wavelengths**.

Constructive interference for a **maxima** or bright fringe occurs when $PD = m\lambda$

$m = 0$ gives central maximum

$m = 1$ gives 1st maximum etc.

i.e. PD = 0, 1λ, 2λ, 3λ

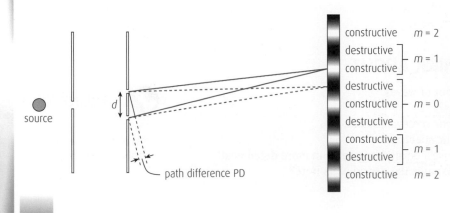

source

d

path difference PD

constructive $m = 2$
destructive
                $m = 1$
constructive
destructive
constructive $m = 0$
destructive
constructive
                $m = 1$
destructive
constructive $m = 2$

Destructive interference takes place when the two light waves are **out of phase**. The waves are out of phase when a crest meets a **trough**.

Destructive interference for a **minima** or dark fringe occurs when

$$\Delta = (m + \tfrac{1}{2})\lambda$$

$m = 0$ gives 1st minimum

$m = 1$ gives 2nd minimum etc.

i.e. $\Delta = \tfrac{1}{2}\lambda,\ 1\tfrac{1}{2}\lambda,\ 2\tfrac{1}{2}\lambda,$

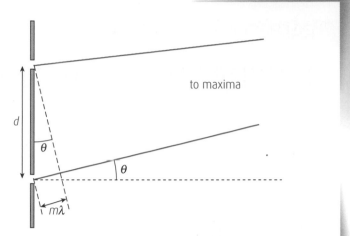

MEASURING THE WAVELENGTH OF LIGHT

The equation $m\lambda = d\sin\theta$ can be used to calculate the wavelengths of light. The slit separation **d** and the angle θ to the mth-order maxima first needs to be measured.

For a maxima, the path difference $= m\lambda$.

Using trigonometry $m\lambda = d\sin\theta$.

THE RANGE OF VISIBLE LIGHT

The **wavelength** of **visible light** ranges from about 4×10^{-7} m to about 7×10^{-7} m.

Here are some spectral colours emitted from cadmium, given in nanometres (10^{-9} m):

blue 480 nm   green 509 nm   red 644 nm

You can often find wavelengths on the exam data sheet.

ONLINE

Learn more about interference at www. brightredbooks.net

THINGS TO DO AND THINK ABOUT

1 You can find interference with many kinds of waves.

2 **Water waves** on the ripple tank show interference.

Note the lines of **rough** (**constructive interference**) and the lines of **calm** (**destructive interference**).

3 **Two loudspeakers** connected to the one signal generator so they are **in phase** produce areas of **louder** and **quieter sound**. Removing one speaker can then increase the sound level in places.

4 You have to watch out for reflections of walls. You can have **interference** between sound **direct** from one speaker and its **reflection** from a side wall.

VIDEO LINK

Head to www. brightredbooks.net and check out the clips.

ONLINE TEST

Be sure to test yourself on this topic at www. brightredbooks.net

DIFFRACTION GRATINGS

A diffraction grating produces a better interference pattern than the basic double slit. Interference patterns are different for monochromatic to white light sources.

THE DIFFRACTION GRATING

A **diffraction grating** consists of a piece of glass into which many thousands of tiny parallel lines have been etched.

The **multiple slits** of the diffraction grating work together to produce a **clearer interference pattern** (with sharp fringes).

As the slits are usually very close together the fringes have greater separation.

Calculating *d*

You are often given the number of **lines per millimetre** and have to work out the **spacing *d*.**

Example:

A grating has 500 lines per millimetre. Calculate the spacing *d* in metres.

Solution:

$$d = \frac{\text{distance}}{\text{number}} = \frac{1 \times 10^{-3}}{500} = 2 \times 10^{-6}\,\text{m}$$

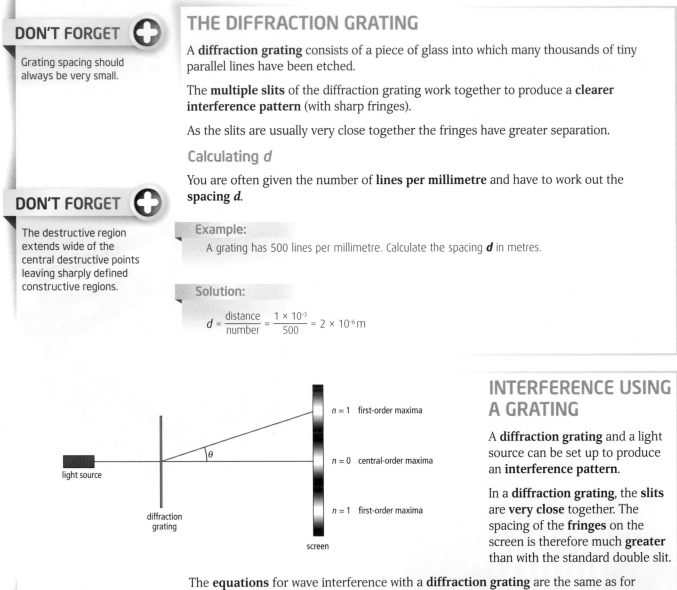

INTERFERENCE USING A GRATING

A **diffraction grating** and a light source can be set up to produce an **interference pattern**.

In a **diffraction grating**, the **slits** are **very close** together. The spacing of the **fringes** on the screen is therefore much **greater** than with the standard double slit.

The **equations** for wave interference with a **diffraction grating** are the same as for Young's slits.

For constructive interference:

$$m\lambda = d\sin\theta$$

Rearranging gives $\sin\theta = \frac{m\lambda}{d}$, so the fringe separation increases with wavelength and as the slits get closer.

The laser

The **laser** is a good light source to use with a diffraction grating. The light from a laser is **monochromatic** (one colour, one frequency) and is **coherent** (same frequency, and all the light has the same phase).

INTERFERENCE OF WHITE LIGHT

When a **white light source** is used with a **diffraction grating**, a **series of spectra** are produced on the screen by **interference**.

A bright **central white maxima** is produced as all the wavelengths of white light interfere **constructively** at the centre.

Red light is **deviated more** than **blue light** since it has the **longest wavelength**. This can be shown from the grating equation

$m\lambda = d\sin\theta$

$\lambda_{red} > \lambda_{blue}$ so $\sin\theta_{red} > \sin\theta_{blue}$ and $\theta_{red} > \theta_{blue}$

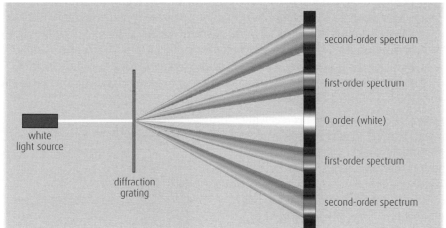

second-order spectrum

first-order spectrum

0 order (white)

first-order spectrum

second-order spectrum

white light source

diffraction grating

DON'T FORGET

Initially, violet on the inside; red on the outside, eventually the colours overlap.

DON'T FORGET

Red shows the largest fringe spacing in the interference pattern, red shows the least deviation when refracting with the triangular prism.

ONLINE

Test your knowledge of diffraction gratings at www.brightredbooks.net

VIDEO LINK

Check out the video about diffraction grating interference at www. brightredbooks.net

ONLINE TEST

Read more about the formula calculation at www. brightredbooks.net

DISPERSION USING A PRISM

A **prism** produces a **single spectrum** by **refraction**.

Red light is deviated **least**, **violet** light is deviated the **most**.

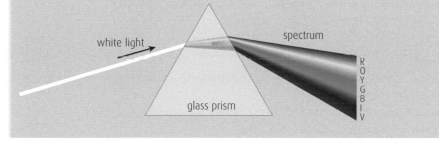

white light

spectrum

glass prism

R O Y G B I V

THINGS TO DO AND THINK ABOUT

1   The tracks of a CD act as a kind of reflection diffraction grating and separate the colours of white light. There are about 625 tracks per millimetre on a CD. This is similar in spacing to many diffraction gratings. How far apart are the tracks?

$d = \dfrac{\text{distance}}{\text{number}} = \dfrac{1 \times 10^{-3}}{625} = 1\cdot6 \times 10^{-6}\,\text{m}$

2   A laser produces a diffraction pattern with a grating with 100 lines per mm. If the first maxima is found at 10°, what is the wavelength and type of radiation?

$d = \dfrac{\text{distance}}{\text{number}} = \dfrac{1 \times 10^{-3}}{100} = 1 \times 10^{-5}\,\text{m}$

$m\lambda = d\sin\theta \Rightarrow 1\lambda = 1 \times 10^{-5}\sin10° \Rightarrow \lambda = 1\cdot74 \times 10^{-6}\,\text{m} = 1740\,\text{nm}$

This is in the **infra-red** region.

THE PHOTOELECTRIC EFFECT

The **photoelectric effect** is the interaction of electromagnetic radiation with electrons in metals, giving rise to the emission of **photelectrons**.

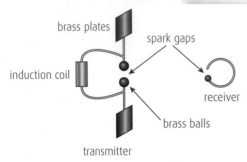

brass plates

spark gaps

induction coil

receiver

brass balls

transmitter

Hertz apparatus

LIGHT AND ELECTRONS

James Clerk Maxwell's theory of electromagnetism, published in 1865, predicted the existence of electromagnetic waves, and that light was such a wave. In 1887 **Heinrich Hertz** was undertaking experiments with electromagnetic waves using a pair of metal spheres. The metal spheres acted as a transmitter, and a thin wire metal loop ending in a metal sphere and a point acted as a receiver. When a spark was generated between the pair of metal spheres, he detected a secondary spark at the receiving loop. This initial work confirmed Maxwell's theory.

In later work, Hertz illuminated his receiving spheres with various colours of light and ultraviolet radiation light (which he obtained from the transmitting spark). He discovered that ultraviolet allowed the biggest spark in his receiver. He also varied the material forming the end of his box. Glass did not allow the ultraviolet through but quartz sheet did.

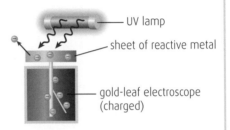

UV lamp

sheet of reactive metal

gold-leaf electroscope (charged)

One year later another German physicist, **Wilhelm Hallwachs**, made a simpler version of Hertz' apparatus to study the effect of shining radiation on charged metal. You could easily do a similar experiment in the physics lab. You might have to investigate the detail more but the main steps are:

- Place a clean, small sheet of zinc metal onto a gold-leaf electroscope.
- Charge the electroscope and zinc metal negatively using a plastic rod.
- The charges spread out and the gold leaf rises as like charges repel.
- Hold an ultraviolet lamp over the charged zinc.
- The leaf will fall as the metal discharges.

We find the following:

- We have to know how to charge the metal negatively. If it is charged positively, the radiation has no effect.
- When a visible colour such as red, green or blue light there is no effect.
- When white light is used, there is no effect.
- When intense white light is used there is still no effect.
- Weak ultraviolet causes a slow discharge, intense ultraviolet causes a rapid discharge.

Hertz and Hallwachs reported their observations but the effect was not understood until 1899, when **J.J. Thomson** identified that the **ultraviolet** was causing **electrons** to be emitted from the surface of the metal, the same particles as in cathode rays.

PHOTOELECTRIC EFFECT

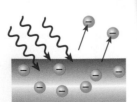

We learn from the Hertz and Hallwachs experiments that electromagnetic radiation and electrons can interact. When certain **metals** are exposed to **high frequencies** of electromagnetic **radiation** (such as **ultra-violet** light), **electrons** are **ejected** from the surface.

This interaction between light and electrons is called the **photoelectric effect**.

There is a minimum **threshold frequency** f_0 for this effect to occur.

PHOTOELECTRIC CURRENT

The electrons ejected from the surface of a metal by electromagnetic radiation can be called **photoelectrons**.

By 1902, **Philipp Lenard**, an assistant of Hertz, had published more detailed results of his experiments into the photoelectric effect. (A Hungarian/German his studies were awarded with a Nobel Prize. He went on to develop Nazi sympathies. You may find this useful for your interdisciplinary studies.)

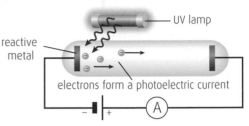

The metal plate is enclosed in a vacuum. Photoelectrons ejected from the plate can be accelerated through the electric field between the plates and cross the vacuum to form a photoelectric current.

The results are consistent with our previous findings:

- **Electrons** are **ejected**. (Reactive metals have outer electrons that are easy to eject.)
- There is a minimum **threshold frequency** f_0 for this effect to occur.
- For frequencies **less** than the threshold, increasing the intensity has no effect.
- For frequencies **greater** than the threshold, increasing the irradiance increases the current.

For **frequencies** greater than the **threshold value**, the **photoelectric current** produced by monochromatic radiation is **directly proportional** to the **irradiance** of the **radiation** at the surface.

DON'T FORGET

Understand the results of the photoelectric experiments and why they pose a problem for wave theory.

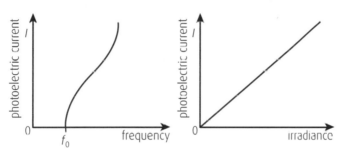

ONLINE

Learn more about the photoelectric current experiment by following the link at www.brightredbooks.net

There is a famous problem with these results. Why cannot low frequency radiation build up enough energy on an electron to eject it?

Light **cannot be a continuous wave** – otherwise the bright, low-frequency wave would build enough energy on an electron to eject it. Also an increase in **irradiance** of a low frequency wave provides more energy and should have an effect.

For classical **wave** theory, the **irradiance** should have an effect and the **frequency** should not. The results say the opposite.

- **Photoelectric emission** from a surface occurs only if the **frequency** of the incident radiation is **greater** than some **threshold frequency** f_0 (which depends on the nature of the surface).
- For **frequencies smaller** than the threshold value, an increase in the **irradiance** of the radiation at the surface will **not** cause **photoelectric emission**.

This gives rise to the next question to answer:

Since wave theory falls down, how does electromagnetic radiation produce this photoelectric effect?

VIDEO LINK

Head to www.brightredbooks.net and watch the clips on this topic.

ONLINE TEST

Test your knowledge of the photoelectric effect at www.brightredbooks.net

THINGS TO DO AND THINK ABOUT

Find how to charge a gold-leaf electroscope negatively and try out Hallwachs' experiment in the lab.

WAVE-PARTICLE DUALITY

The photoelectric effect is evidence for the particulate nature of light. In this section, we look at the meaning of threshold frequency, work function and calculate the maximum kinetic energy of photoelectrons.

ENERGY OF PHOTONS

The puzzle of why low frequency light could not provide energy to eject electrons from metals requires a complete re-thinking of the nature of light.

In 1905 Albert Einstein proposed the **particulate nature of light**. Studying the **photoelectric effect** he said the light appears to behave as a stream of **particles**. Each high frequency UV particle has enough energy to eject an electron whereas each low-frequency particle (such as from red light) does not. One particle either has enough energy to eject an electron or it fails. Subsequent light particles will do the same.

This was a revolutionary idea at the time. The wave nature of light had considerable evidence before the photoelectric effect caused a problem. Einstein called these particles **light quanta** which carried a **quantum** or fixed amount of energy. We can say light is **quantised**. Later these quanta were called **photons**.

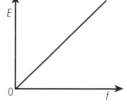

Max Planck had previously discovered that the energy of radiation increased with its frequency. Apply this to Einstein's quanta.

The **energy** of **each** quantum of light, or **photon**, is given from:

$$E \propto f \text{ so } \quad E = hf \quad h \text{ is called Planck's constant.}$$

h is the ratio $\frac{E}{f}$. As frequency increases, energy increases, so h is a constant.
$h = 6\cdot63 \times 10^{-34}\,\text{Js}$.

This constant shows up in many different areas of quantum mechanics.
Also $f = \frac{c}{\lambda}$ so $E = h\frac{c}{\lambda}$ where $c = 3\cdot00 \times 10^8\,\text{ms}^{-1}$.

Examples:

1   What is the energy of a photon of red light of frequency $4\cdot55 \times 10^{14}\,\text{Hz}$?
2   What is the energy of a photon of blue light of wavelength $480\,\text{nm}$?

Solutions:

1   $E = hf = 6\cdot63 \times 10^{-34} \times 4\cdot55 \times 10^{14} = 3\cdot02 \times 10^{-19}\,\text{J}$
2   $E = h\frac{c}{\lambda} = 6\cdot63 \times 10^{-34} \times \left(\frac{3\cdot00 \times 10^8}{4\cdot80 \times 10^{-7}}\right) = 4\cdot14 \times 10^{-19}\,\text{J}$

THRESHOLD FREQUENCY AND WORK FUNCTION

Einstein understood the **work function** w_0 as the amount of **energy** an electron needs to absorb in order to be released from its metal atom. In the photoelectric effect, an **electron** will absorb the energy of a **single photon**. If the energy of the photon is **greater** than the **work function**, the electron will be **ejected**. The minimum frequency of a **photon**, which will have this energy, is the **threshold frequency** f_0.

Threshold frequency = f_0, the minimum **frequency** of a photon to cause **photoemission**.

Work function = w_0, the minimum **energy** to cause **photoemission** from a certain metal.

The minimum energy of a **photon** will have this energy, $E_0 = hf_0$.

light photons

electrons ejected from the surface

sodium metal

UV photon

photoelectron

DON'T FORGET

A **bright** light source produces **more photons** per second causing **more electrons** to be **ejected**, but only if the **frequency** is **high enough**.

DON'T FORGET

The work function and threshold frequency are different for different metals.

DON'T FORGET

A photon of blue light has more energy than a photon of red light.

WAVE-PARTICLE DUALITY

Interference experiments suggest a **wave** nature of light, but the **photoelectric effect** suggests a **particle** nature of light. Both natures exist, and this is called **wave-particle duality**.

A beam of **radiation** can be regarded as a stream of **individual energy bundles** called **photons**, each having an **energy** dependent on the **frequency** of the radiation.

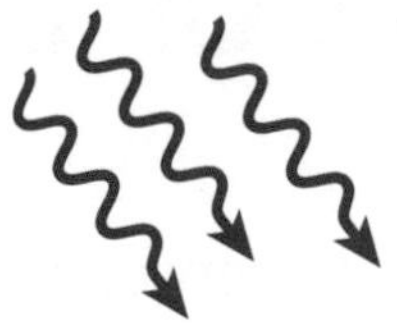

Light beam, waves or particles?

KINETIC ENERGY OF PHOTOELECTRONS

A **photon** with a **frequency higher** than the **threshold frequency** will have **energy** which is **greater** than the **work function** of the metal. The spare or **excess** energy supplies the electron with **kinetic energy**.

Incoming photon energy = work function + kinetic energy for photoelectron:

$$E = w_0 + E_k \qquad\qquad hf = hf_0 + \frac{1}{2}mv^2$$

Photoelectrons are ejected with a **maximum kinetic energy**, given by the difference between the **energy** of the **incident photon** and the **work function** of the **surface.**

Kinetic energy of the electron: $E_k = hf - hf_0$

Example:

Invisible light of wavelength 300 nm is incident on a clean aluminium surface and causes photoemission. The work function of the aluminium is 6.53×10^{-19} J. Calculate the maximum kinetic energy of the photoelectrons.

Solution:

First, work out the frequency: $c = f\lambda$
$3.00 \times 10^8 = f \times 3.00 \times 10^{-7}$
$f = 1 \times 10^{15}$ Hz
Use this frequency in the equation for kinetic energy
$E_k = hf - hf_0$
$E_k = (6.63 \times 10^{-34} \times 1 \times 10^{15}) - (6.53 \times 10^{-19})$
$E_k = 6.63 \times 10^{-19} - 6.53 \times 10^{-19}$
$E_k = 1.00 \times 10^{-20}$ J

Notice that this invisible light is ultraviolet. However at this wavelength and frequency the energy of the photons is only slightly more than the work function. The energy though is enough to eject electrons from the aluminium.

What would happen if the irradiance was trebled? The number of photons would treble and as each photon can eject one photoelectron, the number of photoelectrons ejected would also treble.

Stopping potential

Electrons are emitted with **kinetic energy**. An electric field is applied against the emitted photoelectrons. A **voltage** is recorded which stops the electrons fully crossing the tube. The current stops.

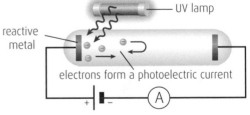

UV lamp

reactive metal

electrons form a photoelectric current

Example:

The stopping potential for electrons during photoemission is 2 V. Calculate the maximum velocity at which these electrons are ejected.

Solution:

$qV = \frac{1}{2}mv^2$
$1.6 \times 10^{-19} \times 2 = \frac{1}{2} \times 9.1 \times 10^{-31} \times v^2$
$v = 8.4 \times 10^5$ ms^{-1}.

ONLINE

Learn more about the photoelectric effect at www.brightredbooks.net

DON'T FORGET

Radiation: we see its wave nature in some experiments and particle nature in other experiments.

DON'T FORGET

Increasing irradiance increases the number of photoelectrons. Increasing frequency increases the kinetic energy.

DON'T FORGET

E_k gives the maximum kinetic energy as the work function gives the minimum energy required.

VIDEO LINK

Watch the clips at www.brightredbooks.net

ONLINE TEST

Head to www.brightredbooks.net to test yourself on this topic.

THINGS TO DO AND THINK ABOUT

Find out why Robert Millikan tried to disprove Einstein's idea and why they were both awarded a Nobel Prize. Find out more about Nobel Prize winners in physics around this time.

REFRACTION OF LIGHT

Light is refracted as it passes between materials with different indexes of refraction. The difference between the refractive indexes affects the amount of refraction.

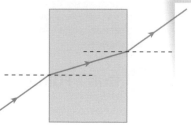

REFRACTION OF LIGHT

When light travels from one medium to another, it undergoes **refraction**.

When light travels from a **less dense to a more dense** material, it **slows down** and bends (or refracts) **towards the normal**. When light travels from a **more dense to a less dense** material, it **speeds up** and bends **away from the normal**. But how much does it bend?

The amount of bending depends on the type of materials used. The way a material bends light is determined by the **refractive index** of the material. The refractive index of a material, with light coming from a vacuum, is known as the **absolute refractive index** of that material. The higher the index, the higher the amount of refraction.

Some **typical refractive indexes** (with yellow light) are:

Diamond 2·42   Crown glass 1·50   Water 1·33   Air 1·00

When **calculating** refractive indices from **practical experiments** to three significant figures, we can use **air** instead of needing to have a **vacuum**.

DON'T FORGET

The angle in a more dense material is always smaller than in a less dense material.

REFRACTIVE INDEX AND SNELL'S LAW

When light travels from medium 1 to medium 2, study the **angles of incidence** θ_1 and **refraction** θ_2 and you find that the relationship between them is not immediately apparent. However, if you plot $\sin\theta_1$ against $\sin\theta_2$, you obtain a **straight line through the origin** showing that $\sin\theta_1$ is **directly proportional** to $\sin\theta_2$. The **gradient** gives the **index**.

The ratio $\frac{\sin\theta_1}{\sin\theta_2}$ is a **constant** when light travels **obliquely** from medium 1 to medium 2.

$\theta_1°$	0	10	20	30	40	50	60	70	80
$\theta_2°$	0	6·5	12·9	19·1	24·8	30	34·5	38	40
$\frac{\sin\theta_1}{\sin\theta_2}$	–	1·53	1·53	1·53	1·53	1·53	1·53	1·53	1·53

The **absolute refractive index**, **n**, of a medium is the **ratio** $\frac{\sin\theta_1}{\sin\theta_2}$ where θ_1 is in a vacuum and θ_2 is in the medium. This gives Snell's law:

$$n = \frac{\sin\theta_1}{\sin\theta_2}$$

In practice, we can use $n = \frac{\sin\theta_{air}}{\sin\theta_{medium}}$

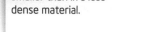
DON'T FORGET

There are **no units** for refractive index.

Example:

A ray of light is shone on the glass block as shown. The ray emerges into air. The angles in the medium and in the air are measured from the normal. θ_2 = 25°, θ_1 = 40°. Calculate the refractive index of the glass.

Solution:

$$n = \frac{\sin\theta_{air}}{\sin\theta_{medium}} = \frac{\sin40}{\sin25} = 1·52$$

The refractive index of this glass is 1·52.

DON'T FORGET

Light ray paths are reversible.

REFRACTIVE INDEX AND VELOCITY

When a **wave** passes from one medium to another with a different refractive index, there is a **change in wave speed**. The **ratio of wave speeds** can be used to find the **refractive index** of a material.

$$n = \frac{V_{air}}{V_{medium}} \text{ or } n = \frac{V_1}{V_2}$$

Examples:

1. A light ray enters a glass block at $3 \times 10^8\,\text{ms}^{-1}$ from air and slows to $2 \times 10^8\,\text{ms}^{-1}$ in the glass.
2. If the refractive index of water is 1·33, what is the speed of light in water?

Solutions:

1. $n = \dfrac{V_{air}}{V_{glass}} = \dfrac{3 \times 10^8}{2 \times 10^8} = 1\cdot50$

2. $n = \dfrac{V_{air}}{V_{water}} \Rightarrow 1\cdot33 = \dfrac{3 \times 10^8}{V_{water}} \Rightarrow V_{water} = 2\cdot26 \times 10^8\,\text{ms}^{-1}$

REFRACTIVE INDEX AND WAVELENGTH

When a **wave** passes from one medium to another, there is a **change in wavelength**.

The **ratio of wavelengths** can be used to find the **refractive index** of a material.

$$n = \frac{\lambda_{air}}{\lambda_{medium}} \qquad n = \frac{\lambda_1}{\lambda_2}$$

$$f = \frac{v}{\lambda} = \text{constant}$$

As **v decreases, λ decreases.**

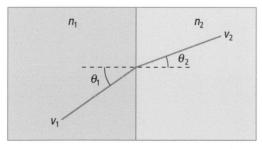

air   glass

λ_1 λ_2

THINGS TO DO AND THINK ABOUT

1. The absolute refractive index of air is 1·0003 to 5 significant figures, so it can be used as the equivalent of the refractive index in a vacuum.

2. Combining equations gives:

   $$n = \frac{\sin\theta_1}{\sin\theta_2} = \frac{v_1}{v_2} = \frac{\lambda_1}{\lambda_2}$$

3. General equations work regardless of direction:

   $$n_1\sin\theta_1 = n_2\sin\theta_2 \qquad n_1 v_1 = n_2 v_2 \qquad n_1 \lambda_1 = n_2 \lambda_2$$

These equations **also** apply between any two materials, such as from water to glass.

$n_{water}\,v_{water} = n_{glass}\,v_{glass}$ and you do not need to worry about direction. When going from a vacuum or air to a medium, use $n_1 = 1\cdot00$.

n_1 n_2

v_2

θ_2

θ_1

v_1

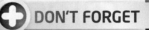

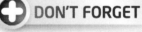

CRITICAL ANGLE, TOTAL INTERNAL REFLECTION AND COLOUR

In this section, we will learn how to find the critical angle when incident light is reflected instead of refracted. We will also look at the conditions for total internal reflection.

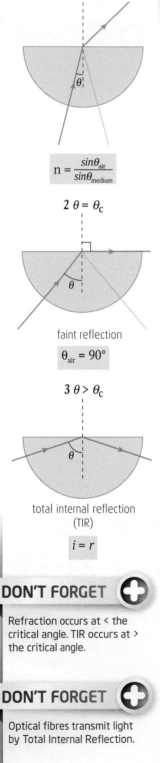

1 $\theta < \theta_c$

$$n = \frac{sin\theta_{air}}{sin\theta_{medium}}$$

2 $\theta = \theta_c$

faint reflection

$\theta_{air} = 90°$

3 $\theta > \theta_c$

total internal reflection (TIR)

$i = r$

CRITICAL ANGLE AND TOTAL INTERNAL REFLECTION

1   When a ray of light passes from an **optically dense material** to a **less optically dense** material, the **refraction** causes the ray to bend **away** from the normal. $\theta_a > \theta_m$

2   When the angle of incidence θ_m is **increased**, there will come a point where the angle of refraction will reach **90°**. The angle of incidence, which causes an **angle of refraction equal to 90°**, is called the **critical angle** θ_c.

3   With an **angle of incidence greater** than the **critical angle**, all the light energy is **reflected** and **none** is refracted. This is **total internal reflection (TIR)**.

Critical angle

At the **critical angle**,

- the angle in the medium = the critical angle, $\theta_m = \theta_c$

- the angle in air is 90°, $\theta_{air} = 90°$

- $n = \frac{sin\theta_{air}}{sin\theta_{medium}} = \frac{sin90}{sin\theta_c} = \frac{1}{sin\theta_c}$ $\boxed{n = \frac{1}{sin\theta_c}}$ or $\boxed{sin\,\theta_c = \frac{1}{n}}$

Examples:

1   Calculate the critical angle for diamond, which has refractive index = 2·42.
2   What is the refractive index of a material whose critical angle is 48°?
3   What happens to a ray incident at 55° on a glass-to-air surface? (n_{glass} = 1·50)?

Solutions:

1 $sin\theta_c = \frac{1}{n} = \frac{1}{2·42} = 0.413$ Critical angle, $\theta_c = 24·4°$

2 $n = \frac{1}{sin\theta_c} = \frac{1}{sin48} = 1.35$ Refractive index, $n = 1·35$

3 $n = \frac{sin\theta_{air}}{sin\theta_{médium}}$ $1.50 = \frac{sin\theta_{air}}{sin55}$

When you use a calculator for this calculation, you will get an error for θ_{air}

This error is occurring because the equation is trying to have $sin\theta_{air} > 1$ which is not possible. An impossible physical situation is being implied. The light cannot refract out greater than 90°. In fact, refraction is not occurring, and this equation no longer applies. Total internal reflection is occurring, and the angles are equal:

$\Rightarrow$ TIR occurs, $\boxed{i = r} = 55°$.

DON'T FORGET ✚

Refraction occurs at < the critical angle. TIR occurs at > the critical angle.

DON'T FORGET ✚

Optical fibres transmit light by Total Internal Reflection.

COLOUR

White light contains a **spectrum** of frequencies.

Higher-frequency blue rays are **refracted more** than the **lower-frequency red rays**.

The **refractive index** depends on the **frequency** of the incident light.

$$n = \frac{sin\theta_{white}}{sin\theta_{red}} \qquad n = \frac{sin\theta_{white}}{sin\theta_{blue}}$$

Thus the **refractive index** of a material may **vary** depending on the **colour or frequency** of light in use. (The variation could be from 1·51 to 1·53 in glass.) Note $n_{violet} > n_{red}$ thus violet bends more than red.

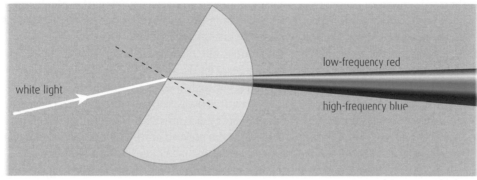

white light

low-frequency red

high-frequency blue

DON'T FORGET

'**Dispersion**' is due to **different refractive indices** for **different colours**.

ONLINE

Read more about refraction at www.brightredbooks.net

VIDEO LINK

Check out the clips at www.brightredbooks.net

ONLINE TEST

Head to www.brightredbooks.net to test yourself on this topic.

THINGS TO DO AND THINK ABOUT

1   There are many options for equations:

$$n = \frac{sin\theta_1}{sin\theta_2} = \frac{sin\theta_{air}}{sin\theta_{medium}} = \frac{v_1}{v_2} = \frac{\lambda_1}{\lambda_2} = \frac{1}{sin\theta_c}$$

2   How would you calculate the **range of dispersion** (angle between red light and blue light) above? Refer to the diagram above.

3   **Optical fibres** carry signals for **telecommunications**. They rely on **total internal reflection**, and the signal travels a long distance. Normally **monochromatic light** is used. As the **different colours have a different refractive index**, this means they would travel at slightly **different speeds**. The information would not stay together and the communication would be lost.

BOHR MODEL AND SPECTRA

In this section, we will be learning the Bohr model of the atom, the terms **ground state**, **energy levels**, **ionisation** and **zero potential energy**. We will also study line and continuous emission spectra, absorption spectra and energy level transitions.

excited states

ground state

nucleus

ENERGY LEVELS

Rutherford's model of the atom was a **central positive charge** with **negative electrons** around. Niels **Bohr** used the ideas of Einstein and Planck to extend this model to suggest that **electrons** in a free atom occupy **discrete energy levels**.

- Electrons are placed in circular orbits around the nucleus.
- There are a limited number of allowed energy levels, so the orbitals are quantised.
- An electron, bound in an atom, occupies certain states, equal to the allowed levels. An electron may gain energy and jump to an excited state. The electron may then drop from an excited state to a lower state, emitting energy as a quantum or photon of radiation.
- An electron can be bound to the nucleus in the **ground state** or a number of **excited** states. All these energy levels for bound electrons have **negative values**.
- An electron in the ground state has the least energy. The ground-state energy level is a measure of the energy needed to **unbind** or **ionise** the electron.
- An electron just freed from the atom is then said to have reached zero potential energy value.
- Once an electron is free from an atom, it gains positive kinetic energy.

DON'T FORGET

Not all the energy levels are shown. The higher levels get closer together.

E_3 E_4

E_0 E_1 E_2

Energy levels of the hydrogen atom

The hydrogen atom has a single positive proton in the nucleus and a single negative electron.

- The electron has least energy in the ground state, closest to the nucleus.
- An electron moves to a higher energy level when it absorbs the energy of a photon.
- When a photon is absorbed, an electron jumps, or makes a transition, from a lower energy level to a higher energy level.
- The electron can only absorb photons of certain energies exactly matched to the energy difference between two energy levels.
- When an electron makes a transition from a higher energy level to a lower energy level, a photon is emitted.

DON'T FORGET

Remember you can find equations in the Relationships sheet. For Bohr's atom find

$E_2 - E_1 = hf$

DON'T FORGET

Electrons can only make a transition between levels, they cannot land in between.

ionisation level $(E = 0\,J)$

E_4
E_3 $-1\cdot360 \times 10^{-19}\,J$
E_2 $-2\cdot416 \times 10^{-19}\,J$
E_1 $-5\cdot624 \times 10^{-19}\,J$

less energy / more negative

excited states

E_0 $-21\cdot76 \times 10^{-19}\,J$

ground state

$$E_{photon} = hf = E_{excited} - E_{lower}$$

- The frequency of the absorbed or emitted photon can be found from

$$f = \frac{E_{excited} - E_{lower}}{h} \quad \text{or} \quad f = \frac{\Delta E}{h}$$

Example:

A **photon** is **emitted** when an **electron** makes a **transition** from an energy level of $-2\cdot416 \times 10^{-19}\,J$ to a lower level of $-21\cdot76 \times 10^{-19}\,J$. Calculate the frequency of the photon.

Solution:

$$f = \frac{\Delta E}{h} = \frac{-2\cdot416 \times 10^{-19} - (-21\cdot76 \times 10^{-19})}{6\cdot63 \times 10^{-34}} = 2\cdot918 \times 10^{15}\,Hz$$

This is an ultraviolet photon.

DON'T FORGET

An electron absorbs a photon to make an upwards transition. A downwards transition emits a photon.

SPECTRA

The **spectrum** from a light source can be displayed using a **prism** or a **diffraction grating**.

An **emission line** in a **spectrum** will be seen when **electrons** make a **transition**, emitting photons between an **excited energy level** and a **lower energy level**.

The energy to **raise** an electron to an excited state can come from
- a high voltage in discharge tubes
- heat in a filament lamp
- nuclear fusion in the stars.

Line-emission spectrum

Line-emission spectra give an insight into the **structure** of an **atom**. Each **element** has a **different line spectrum**, as each has a **different** structure of **energy levels**.

From four energy levels, six **different frequencies** of photons could be **emitted** and six lines of light can be seen on the **line spectrum**.

$$\Delta E = hf$$

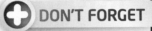

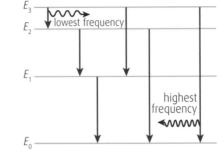

Line-emission spectra are emitted from **low-pressure gases** where there are **free atoms**.

Lines are **brighter** where **more electrons** make a **certain transition**.

Continuous-emission spectrum

Continuous-emission spectra are seen from **solids**, **liquids** and **high-pressure gases**. Their electrons are bonding with other atoms, giving **infinite transitions** and **infinite lines**.

Line-absorption spectrum

The **absorption spectrum** of an element has **black lines** on a **continuous spectrum**, which are in the identical position to the bright lines of that element's line emission spectrum.

An **absorption line** in a spectrum occurs when **electrons** in a lower energy level **absorb radiation** and are excited to a higher energy level.

White light produced from a lamp or the sun has **certain frequencies missing** after passing through a gas cloud. The **gas absorbs** the frequencies of light, which it would normally emit. The gas' electrons absorb the photons whose energies match the energy transitions available. The **black lines** are known as **Fraunhofer lines** after the scientist Joseph von Fraunhofer, who discovered these lines missing from the Sun's spectrum. They are evidence of composition of gases in the Sun's upper atmosphere.

continuous-emission spectrum

line emission spectrum

line absorption spectrum

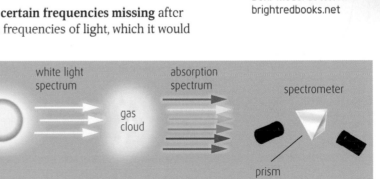

white light spectrum

gas cloud

absorption spectrum

spectrometer

prism

THINGS TO DO AND THINK ABOUT

1. When electrons make a **large energy transition**, the photon frequency will be high. This might lead to **ultraviolet** or even **x-rays** being emitted.
2. After electrons **absorb photons** from **one direction** to create black Fraunhofer lines, they fall back down again and **emit photons**, but this time the photons can be emitted in **all directions** and so are not seen.

IRRADIANCE AND THE LASER

We are now going to learn about irradiance and how to successfully apply the inverse square law.

IRRADIANCE

Irradiance is the term used when electromagnetic **radiation** is incident on a **surface**.

Irradiance is a measure of the **power** incident on a **surface** and is measured in Wm⁻².

DON'T FORGET

A point source is one which emits rays equally in all directions.

Irradiance is the **power per unit area** $I = \frac{P}{A}$

Example:

A 150 W lamp illuminates a screen whose area is 4 m². Calculate the irradiance on the screen.

Solution:

$$I = \frac{P}{A} = \frac{150}{4} = 37 \cdot 5 \, \text{Wm}^{-2}$$

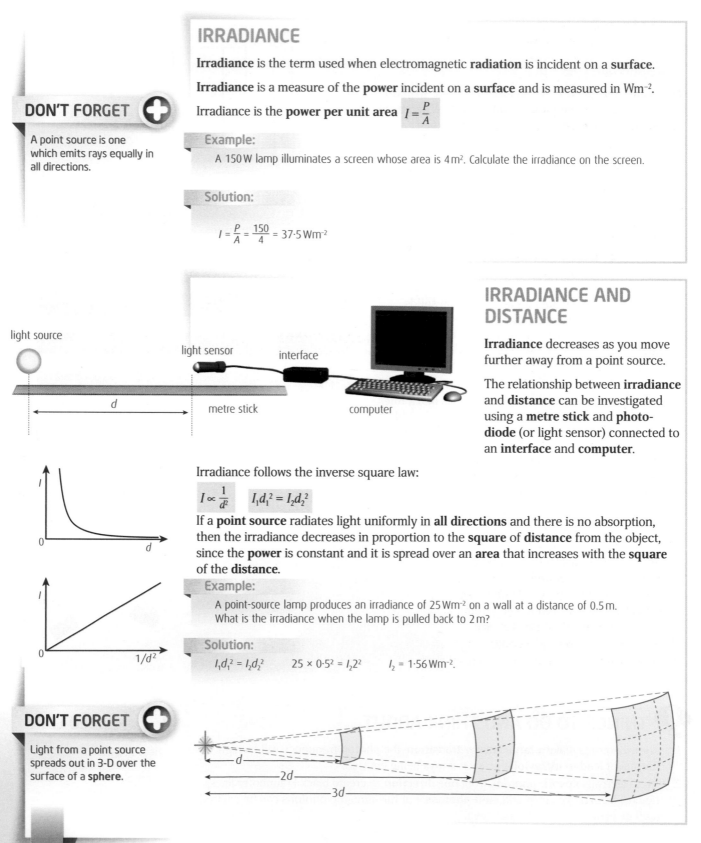

light source

light sensor   interface

d

metre stick   computer

IRRADIANCE AND DISTANCE

Irradiance decreases as you move further away from a point source.

The relationship between **irradiance** and **distance** can be investigated using a **metre stick** and **photo-diode** (or light sensor) connected to an **interface** and **computer**.

Irradiance follows the inverse square law:

$$I \propto \frac{1}{d^2} \qquad I_1 d_1^2 = I_2 d_2^2$$

If a **point source** radiates light uniformly in **all directions** and there is no absorption, then the irradiance decreases in proportion to the **square** of **distance** from the object, since the **power** is constant and it is spread over an **area** that increases with the **square** of the **distance**.

Example:

A point-source lamp produces an irradiance of 25 Wm⁻² on a wall at a distance of 0.5 m. What is the irradiance when the lamp is pulled back to 2 m?

Solution:

$$I_1 d_1^2 = I_2 d_2^2 \qquad 25 \times 0 \cdot 5^2 = I_2 2^2 \qquad I_2 = 1 \cdot 56 \, \text{Wm}^{-2}.$$

I ... d

I ... $1/d^2$

DON'T FORGET

Light from a point source spreads out in 3-D over the surface of a **sphere**.

d
$2d$
$3d$

STIMULATING LIGHT

Spontaneous emission

Light is **emitted** when **electrons** in an **excited state** drop to a **lower energy level**. The drop can be **spontaneous**, happening at a **random time**, and **emitting photons** in **any direction**. This is how most **ordinary sources** make light.

Spontaneous emission of radiation is a **random** process analogous to the **radioactive decay** of a **nucleus**.

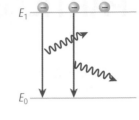

Stimulated absorption

If an **incident photon** is **absorbed**, an electron takes all its **energy**.

The **photon energy** is **absorbed** and the electron will jump to a **higher energy level** if the photon's **frequency** gives it an energy which is an **exact match** for the **difference** in energy levels. $hf = E_2 - E_1$

In normal **crystals** and **gases**, there are **many electrons** in the **lower energy levels**, and it is easy for **photons** to be **absorbed**.

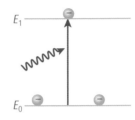

Stimulated emission

When radiation of **energy** hf is incident on an excited atom, the atom may be **stimulated** to emit its **excess energy** hf.

If an electron is in the **excited state**, and the incoming photon's energy matches the difference in energy levels, the photon will **stimulate** the electron to fall and **emit** a photon of **identical frequency**.

In **stimulated emission**, the **incident radiation** and the **emitted radiation** are **in phase** and travel in the **same direction**.

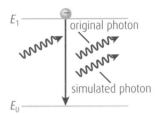

THE LASER

The conditions in the **laser** are such that a light beam **gains more energy** by **stimulated emission** than it loses by absorption – hence Light Amplification by the Stimulated Emission of Radiation. There must be a **population inversion** (more electrons in the **excited state**) in the laser medium. A photon starts a **chain reaction**, as emitted photons become stimulating photons.

The laser has **mirrors** at each end. The photons are **reflected** back and forth creating an **avalanche** effect, and this **amplification** creates a **powerful pulse** of light. Some light is allowed to escape through the **partially reflecting mirror** to create the **laser beam**.

A **beam of laser light** having a **power** even as low as 0·1 mW may cause **eye damage**.

beam

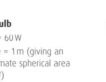

laser

partially reflecting mirror

fully reflecting mirror

Irradiance

School laser	Light bulb
power = 0·5 mW	power = 60 W
beam diameter = 1 mm	distance = 1 m (giving an approximate spherical area of 12 m²)
$I = \dfrac{P}{A} = \dfrac{P}{\pi r^2} = 0 \cdot 5 \times$ $\dfrac{10^{-3}}{3 \cdot 14} \times (0 \cdot 5 \times 10^{-3})^2$ $= 636 \, \text{Wm}^{-2}$	$I = \dfrac{P}{A} = \dfrac{60}{12} = 5 \, \text{Wm}^{-2}$

The diameter of a circular laser beam may stay at approximately 1 mm over a long distance. This means that it has a much higher irradiance than conventional light sources.

DON'T FORGET

Laser light is monochromatic $f = \dfrac{BE}{h}$ coherent, parallel and has **high irradiance**.

ONLINE

Learn more about lasers and the inverse square law at www.brightredbooks.net

VIDEO LINK

Watch the videos about lasers and the inverse square law at www.brightredbooks.net

ONLINE TEST

Head to www.brightredbooks.net to test yourself on this topic.

THINGS TO DO AND THINK ABOUT

1   The power from a point source spreads over a spherical area.

$$I = \frac{P}{A} = \frac{P}{4\pi r^2} \Rightarrow I = \frac{1}{r^2}$$

2   The pupil of your eye has a larger area than the diameter of a laser beam.

point source

r

sphere

PARTICLES AND WAVES QUESTIONS

Practice and revise with help from these examples from the topics you have been studying in this unit.

THE STANDARD MODEL

1 State the order of magnitude that would describe a 100 m race.

2 What quality do particles and antiparticles share?

3 Give a definition of a positron.

4 What do fermions consist of?

5 Hadrons are composite particles. What are they made of?

6 How many quarks do baryons and mesons have?

7 What is the charge on an up and a down quark?

8 State what quarks a proton has and derive its charge from these.

9 What are the force mediating particles of the Standard Model?

10 Which particle was predicted by a theoretical physicist from Edinburgh but not discovered for nearly 50 years?

FORCES ON CHARGED PARTICLES

11 What does an electric charge experience in an electric field?

12 What happens when an electric field is applied to a conductor?

13 What has to be done to move charge in an electric field?

14 Define potential difference.

15 Define the volt.

16 Calculate the potential difference of a battery supplying 18 J of energy to 2 C of charge.

17 An electron is accelerated through a potential difference of 1000 V. Calculate the velocity it will reach.

18 State Oersted's discovery.

19 Describe the direction that a force will exert when a charge crosses a magnetic field.

20 What type of particle accelerator has the world's longest path?

NUCLEAR REACTIONS

21 Describe Rutherford's model and how he derived it.

22 What is meant by alpha, beta and gamma decay of radionuclides?

23 Explain nuclear fusion.

24 Identify the processes occurring in nuclear reactions written in symbolic form.

25 Explain nuclear fission.

INTERFERENCE AND DIFFRACTION

26 Describe how the frequency of a wave depends on its source.

27 What is considered the test for a wave?

28 State at least four behaviours of waves.

29 The path difference to the second order maxima in an interference pattern is 1000 nm. Calculate the wavelength of the radiation.

30 State approximate values for the wavelengths of red, green and blue light.

31 What does the energy of a wave depends on?

32 What is the meaning of the term *in phase*?

33 What is the meaning of the term *coherent*?

34 Describe the principles of a method for measuring the wavelength of a monochromatic light source, using a grating.

WAVE PARTICLE DUALITY

35 What are the conditions required for photoelectric emission to occur?

36 What is the relationship between photoelectric current and irradiance?

37 Describe the nature of a beam of radiation.

38 Calculate the energy of a photon of wavelength 500 nm.

39 State how to calculate the maximum kinetic energy of photoelectrons.

REFRACTION OF LIGHT

40 State Snell's law.

41 A ray of light enters a glass block at 42° from the normal. If the angle of refraction in the glass is 22°, calculate the refractive index of this glass.

42 Calculate the velocity that the ray in Q41 will travel at in the glass.

43 Explain what is meant by the *critical angle* θ_c.

44 State why a medium such as glass can have a range of refractive indexes.

45 Describe the meaning of *total internal reflection*.

SPECTRA

46 Explain what is meant by irradiance.

47 Where are the electrons in a free atom?

48 Explain the occurrence of absorption lines in the spectrum of sunlight.

49 Describe what spontaneous emission of radiation is.

50 Explain the function of the mirrors in a laser.

💭 THINGS TO DO AND THINK ABOUT

Check with the previous pages to ensure you can answer these correctly and to reinforce your learning.

COURSE ASSESSMENT

HOW TO TACKLE THE ASSIGNMENT 1

The assignment requires you to carry out an investigation into a physics topic at a level of demand appropriate to CfE Higher Physics. Experimental work is carried out in class and a report on your investigation must be produced under exam conditions.

Your teacher should suggest appropriate topics to investigate and offer reasonable advice on how to carry out the investigation.

STAGES OF THE ASSIGNMENT

The assignment has two stages:

- a research stage
- a communication stage

Research

During the **research stage**, you will gather information or data from scientific journals or internet sites and also from experimental activity you have carried out in school. A minimum of two sources of information must be gathered and one of these must be your practical experiment.

Communication

The **communication stage** is the production of your report. Your report will be written under exam conditions and should contain about 800–1500 words, excluding tables, charts and diagrams. The report is then sent to SQA for marking where a total of 20 marks is available.

MARK ALLOCATION

As mentioned above, your report is worth up to 20 marks (out of 120 total marks) towards your final grade. The table below shows how these 20 marks are allocated.

Full details of how these nine categories are marked should be provided by your teacher. Some **additional** advice on the preparation of your assignment report is given here.

Category	Mark allocation
Aim(s)	1
Applying knowledge and understanding of physics	4
Selecting information	2
Processing and presenting data/information	4
Uncertainties	1
Analysing data/information	2
Conclusion(s)	1
Evaluation	3
Presentation	2

Aim (1 mark)

The aim must clearly describe what is being investigated. An aim that is a bit vague will not be awarded the mark. Let's look at examples of aim statements which are too vague with better alternatives.

> **Example:**
> ✗ To investigate the effectiveness of sun cream. (vague)
> ✓ To investigate the absorption of UV light by different sun creams. (better)

> **Example:**
> ✗ To investigate the amount of light passing through an optical fibre. (vague)
> ✓ To investigate how the amount of bending of an optical fibre affects the transmission of light through the fibre. (better)

Example:

- (✗) To investigate different properties of LEDs. (too vague – which properties?)
- (✓) To investigate how the switch on voltage of an LED changes with the colour of the LED (or wavelength of light emitted by the LED) (much better)

Make sure you have a properly thought out aim before the final write up. This should ensure you achieve the "aim" mark.

Apply knowledge and understanding of physics (4 marks)

A convenient sub-heading for this category is **Underlying Physics**. The physics theory including equations (relationships) should be described here. The investigation should be at Higher physics level so the underlying theory should be familiar to you and you will have to write this up at the report stage under exam conditions.

A bit of historical information is always interesting but will not count towards scoring marks in this category. Similarly several paragraphs on the biological effects and dangers of sunburn will not score any marks unless there is some physics described.

Advice for some popular topics is given in the following table:

Investigation topic	Possible physics development
exoplanets investigations	inverse square law; $I \propto \dfrac{1}{d^2}$; $I = \dfrac{P}{A}$
sunscreen investigations	UVA, UVB, UVC; photons; $E = hf$; $c = f\lambda$;
LED investigations	semiconductor theory; $E = hf$; $c = f\lambda$;
seismology investigations	longitudinal/transverse waves; friction; E_p; E_K

DON'T FORGET

The inverse square law is or $I \propto \dfrac{1}{d^2}$; $I = \dfrac{k}{d^2}$ not $= \dfrac{1}{d^2}$

A report scoring full marks in the underlying physics category will probably develop several physics principles and relationships. An occasional mistake need not prevent you from scoring full marks.

Selecting information (2 marks)

The raw data gathered in the researching stage must be included in your report. This data must be from a minimum of two sources and you are allowed to bring this data with you to the write up stage. Your experimental results will be one of the sources of data. The second source will be from your research and may be a printed page from a website.

Both sets of raw data must be relevant to your aim and must be sufficient to reach a conclusion.

If your experimental results lead to a relevant conclusion but your downloaded internet data is insufficient or does not match the aim then you will only receive 1 mark.

THINGS TO DO AND THINK ABOUT

Read through the advice given here carefully. Make a checklist for yourself to ensure that your report covers all of the bases outlined in these pages.

HOW TO TACKLE THE ASSIGNMENT 2

MARK ALLOCATION (CONTD.)

Processing and presenting data/information (4 marks)

You must produce a graph from your experimental results to support your aim. The graph will be marked as follows:

- All headings/labels/units/scales are correct **(1 mark)**. If any of these is missing, such as a missing unit, then 0 marks will be awarded here.
- (Almost) All points are plotted correctly **(1 mark)**. If less than 90% of points are correctly plotted then 0 marks will be awarded here. A poor choice of scale making it difficult to check the accuracy of plotting will score 0 marks. If you use a graphing package like Microsoft Excel then you must include minor gridlines otherwise 0 marks here.
- The graph must have a line (or curve) of best fit **(1 mark)**. A straight line joining the first and last point is not necessarily the line of best fit. Joining the points with a series of straight lines is not the line of best fit.

The second source of data (such as internet data) has to be processed as well and it too may produce a graph. Alternatively it may be processed into an extra column in a table. All headings/units and 90% of calculations must be correct. A sample calculation must be shown.

The final mark in this category is awarded for cross-referencing your two sources. The full URL must be given if an internet site provided one of your sources. Your experiment must have a title and aim so it can be cross-referenced.

Most reports will require an experimental title/aim **and** a full URL. **(1 mark)**

Uncertainties (1 mark)

Appropriate reading and random uncertainties should be included in your report.

You should include a statement like '*the reading uncertainty of each voltage is ± 0·01 V*' for each different measurement you make in your experimental results.

There must be a sample calculation of random uncertainty somewhere in your report.

> **Example:**
>
> $$\text{random uncertainty} = \frac{max - min}{n}$$
> $$= \frac{3 \cdot 7 - 3 \cdot 2}{5}$$
> $$= \pm 0 \cdot 1 \ N$$

DON'T FORGET

URL stands for Uniform Resource Locator.

DON'T FORGET

Remember to include the unit.

Analysing data/information (2 marks)

Analysis may include comparisons; trends; proving direct proportion between two variables; using the gradient of a straight line graph to calculate a constant; or comparing the experimental value with the accepted value.

Some correct analysis should score 1 mark with a more complete analysis scoring 2 marks.

Linking your experimental results to the data processed from your internet research is necessary to achieve 2 marks.

Conclusion (1 mark)

Your conclusion must relate to your aim and must be supported by your experimental data.

Example:

The transmission of light through an optical fibre decreased with the amount of bending of the fibre.

Evaluation (3 marks)

Have a separate heading for evaluation and list at least three pieces of evaluation. Each evaluation statement should include an appropriate justification.

Example:

- *Dark clothing should have been worn to reduce reflection of light on to the lightmeter.*
- *The light bulb ideally should be much smaller to resemble a point source of light.*
- *The internet data is reliable as NASA is a reputable government agency.*

Try to avoid variations of the following statement: '*my results would have been better if I had better equipment*'. This has become something of an 'old chestnut' among SQA markers and will not score any marks. Evaluation statements demand a little more sophistication than this.

Three valid evaluation statements will score 3 marks (one mark for each statement)

Presentation (2 marks)

The first mark is for having an appropriate title at the start of your report **and** having the references page at the end of the report.

Note that 'Higher Physics Assignment' is not an appropriate title. The title should refer to what your research is about.

The references page must be at the end unless any following pages are clearly labelled 'Appendices'.

Good advice might be to have no appendices and make sure that the references page is last.

The second mark in the presentation category is for listing your references on the references page. Your two sources must be among the references in the references page. Your experimental work should be listed with the title of your experiment and aim, and the internet research is listed with its full URL. Note that a shortened version like www.bbc.co.uk is not a full URL.

Example:

- *source 1.   Title: ****************
- *source 2.   Aim: ****************
- *reference 3   insert here if there are more than two references*

🗨 THINGS TO DO AND THINK ABOUT

Some students will choose scales for graphs similar to those on the following graph.

Can you easily identify the coordinates of point X?

Now try plotting the point (5·5 mA, 3·6 V).

Is this a good choice of scales?

Always better to have major gridlines increasing in units of 1, 2 or 5.

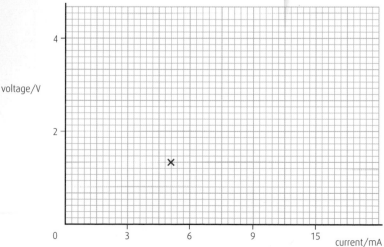

GRAPHICAL ANALYSIS

One of the questions in the CfE Higher Physics exam will look unfamiliar to you and will be based on content which is not in the syllabus. The physics relationship involved will be stated in the question and your skill at handling this unfamiliar situation will be assessed. The question will probably include a table of raw data from which a graph will be drawn and a physical quantity determined using the graph.

DON'T FORGET

Your points must be on the graph line and not from the table of data

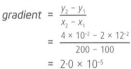

$$gradient = \frac{y_2 - y_1}{x_2 - x_1}$$
$$= \frac{4 \times 10^{-2} - 2 \times 12^{-2}}{200 - 100}$$
$$= 2 \cdot 0 \times 10^{-5}$$

DON'T FORGET

The gradient unit will be mT × 10⁻³ but this is not the skill being assessed here.

DON'T FORGET

$r = \frac{mv}{Q} \times \frac{1}{B}$

is like $y = mx$ where m is the gradient.

$gradient = \frac{mv}{Q}$

$v = \frac{gradient \times Q}{m}$

$= \frac{2 \times 10^{-5} \times 1 \cdot 6 \times 10^{-19}}{m}$

$= 3 \cdot 5 \times 10^{6}\,m\,s^{-1}$

Example:

A beam of electrons follows a circular path in a uniform magnetic field.

The radius, r, of the semicircular path depends on the magnetic field strength B.

The unit of magnetic field strength is the tesla T.

A student measures the radius of the semicircular path for several different values of magnetic field strength and the results are shown in the table.

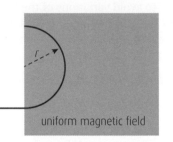

electron beam → uniform magnetic field

(a) Using squared ruled paper, plot a graph of r against $\frac{1}{B}$.

(b) Calculate the gradient of your graph. The gradient unit is not required.

(c) The radius r of the semicircular path is given by the relationship

$$r = \frac{mv}{Q} \times \frac{1}{B}$$

where r is the radius of the semicircular path in metres
m is the mass of the electron in kilograms
v is the speed of the electrons in metres per second
B is the magnetic field strength in tesla
Q is the charge on an electron in coulombs.

B/T	r/m	$\frac{1}{B}$/T⁻¹
$4 \cdot 0 \times 10^{-3}$	$4 \cdot 9 \times 10^{-3}$	250
$5 \cdot 0 \times 10^{-3}$	$4 \cdot 2 \times 10^{-3}$	200
$8 \cdot 0 \times 10^{-3}$	$2 \cdot 6 \times 10^{-3}$	125
$10 \cdot 0 \times 10^{-3}$	$2 \cdot 1 \times 10^{-3}$	100
$20 \cdot 0 \times 10^{-3}$	$0 \cdot 9 \times 10^{-3}$	50

Use this relationship and the gradient of your graph to calculate the speed of the electron in the magnetic field.

Solution:

(a) Plotting r against $\frac{1}{B}$ should have r on the vertical axis. Think 'plotting y against x' (it would be unusual to hear 'plot x against y').

Make sure you label each axis and insert the correct unit on each axis.

Plotting the points should be straightforward if your choice of scale is good.

The best straight line on this graph should have points on either side of the line. Joining the first point to the last point will be incorrect as all the other points will be above this line.

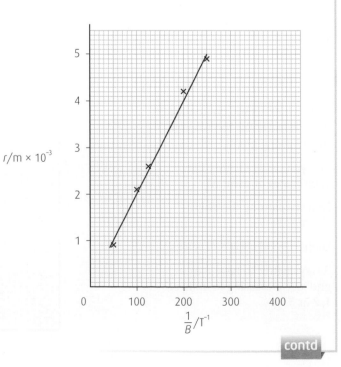

r/m × 10⁻³

$\frac{1}{B}$/T⁻¹

contd

Solution (continued):

A 12-inch ruler is better than a 6-inch ruler for graph drawing. Using the edge of a protractor will possibly give two slightly different straight lines and be penalised.

(b)   Select two points **on your graph**, for example (100, 2 × 10⁻³) and (200, 4 × 10⁻³).

(c) $r = \dfrac{mv}{Q} \times \dfrac{1}{B}$

You must use your value for the gradient in this question. Simply substituting values for r, B, Q and m into the relationship $r = \dfrac{mv}{Q} \times \dfrac{1}{B}$ is not answering the question or demonstrating the data handling skill being assessed here.

The relationship could have been written as $r = \dfrac{mv}{QB}$ but this would have disguised the fact that the gradient of the graph is $\dfrac{mv}{Q}$.

⚙ EXERCISE:

1   The period of a pendulum T varies with the length l of the pendulum and the relationship is

$$T^2 = \frac{4\pi^2}{g} \times l$$

The value of T is measured for various values of l and the results for T^2 and l are in the following table.

l/m	0·29	0·64	0·76	1·0	1·4
T^2/s²	0·75	1·80	2·25	3·10	4·8

(a) Plot a graph of T^2 against l.

(b) Show that the gradient is approximately 3·75. The unit is s²m⁻¹.

(c) Use the relationship above and your value of gradient to calculate the value of g.

☁ THINGS TO DO AND THINK ABOUT

The data being graphed could give a curved line rather than a straight line. Draw the best curved line you can and **not** a series of straight lines from point to point. This is a common mistake made by students.

Another common mistake made in exams is drawing a series of curved lines rather than a single line of best fit, for example:

Best practice would be to draw graph lines in pencil and rub out as required to avoid losing marks.

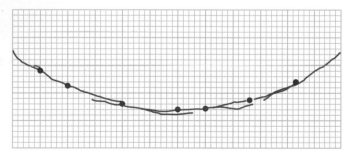

APPENDICES

ANSWERS

OUR DYNAMIC UNIVERSE

1 Displacement is the distance travelled in a stated direction from the starting point.

2 Velocity is the rate of change of displacement.

3 A vector quantity is defined by both its magnitude and its direction.

4 361 m at 34° S of E

5 Acceleration and displacement

6 $36\,\text{ms}^{-1}$, $17\,\text{ms}^{-1}$, 3·7 s and 63 m to 2 sig. figs. $g = 9\cdot8\,\text{ms}^{-2}$ and no friction.

7 Acceleration is the rate of change of velocity.

8 See pp. 18–19

9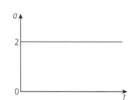

10 Constantly increasing velocity, constant acceleration

11 See pp. 20–21

12 $14\,\text{ms}^{-1}$

13 The unit of force is the Newton. 1 N is defined as the resultant force, which causes a mass of 1 kg to accelerate at $1\,\text{ms}^{-2}$.

14 $140\cdot2\,\text{ms}^{-2} \Rightarrow 140\,\text{ms}^{-2}$ (sig. figs!)

15 424 N

16 Momentum is the product of mass and velocity.

17 The law of conservation of linear momentum can be applied to the interaction of two objects moving in one direction, in the absence of net external forces.

18 An elastic collision is one where both momentum and kinetic energy are conserved.

19 In an inelastic collision momentum is conserved but kinetic energy is not.

20 $2\,\text{ms}^{-1}$ in original direction

21 $2\,\text{ms}^{-1}$ in opposite direction to bullet

22 $E_p = E_W + E_k$ which is $mg(h_2 - h_1) = Fd + \frac{1}{2}mv^2$

23 Impulse = force × time

24 Impulse = change of momentum

25 Ns and kgms^{-2}

26 $16\,000\,\text{kgms}^{-2}$ and $40\,000\,\text{N}$, less force but for a longer time.

27 The force exerted on unit mass.

28 Time to impact = 1·01 s; Vertical velocity = $9\cdot899\,\text{ms}^{-1}$; Range of impact = 35·35 m; Angle from horizontal = 8°.

29 Assume your mass is 70 kg, W = 259 N.

30 The gravitational force varies inversely with distance between two objects, squared.

31 $7 \times 10^{-8}\,\text{N}$, pupils are round, small or uniform density.

32 Both measure $c = 3 \times 10^8\,\text{ms}^{-1}$.

33 0·625 => 0·6 years

34 30·3 m.

35 1316 Hz then 1200 Hz then 1103 Hz. 0·26 m.

36 Doppler shift

37 480 ls, $1\cdot5 \times 10^{-5}\,\text{ly}$

38 100 000 ly

39 −0·0737, $2\cdot21 \times 10^7\,\text{ms}^{-1}$, moving away

40 −0·09

41 $2\cdot21 \times 10^7\,\text{ms}^{-1}$

42 The peak wavelength decreases, the energy increases.

ELECTRICITY

1 $f = \frac{1}{T}$ ... see p 58

2 10 V

3 3·5 A

4 9 V

5 540 C

6 48 W

7 See p 61

8 See p 61

9 12 V and 24 V

10 12 Ω

11 For an initially balanced Wheatstone bridge, as the value of one resistor is changed by a small amount, the out-of-balance p.d. is directly proportional to the change in resistance.

12 The e.m.f. of a source is the electrical potential energy supplied to each coulomb of charge which passes through the source.

13 An electrical source is equivalent to a source of e.m.f. with a resistor in series, the internal resistance.

14 See p 65

15 See p 65

16 The charge on two parallel conducting plates is directly proportional to the p.d. between the plates.

17 Capacitance is the ratio of charge to p.d.

18 The unit of capacitance is the farad and that 1 farad is 1 coulomb per volt.

19 7000 μF

20 See p 68

21 The work done to charge a capacitor is given by the area under the graph of charge against p.d.

22 $E = \frac{1}{2}QV$, $E = \frac{1}{2}CV^2$, $E = \frac{1}{2}Q^2/C$

23 $Q = 4400\,\text{pC}$. $E = 22\,000\,\text{pJ}$.

24 See p 68

25 $I_{max} = 0.018\,\text{A}$ initially. V = 9 V finally.

26 Storing energy, storing charge, smoothing voltage, filters, blocking d.c. while passing a.c., strobe flash, touch screens

27 In energy levels.

28 In energy bands.

29 The bands overlap or the valence band is the conduction band as it is only part full.

30 The bands are far apart from each other.

31 The gap is very narrow.

32 Pure semiconductor vs doped semiconductor.

33 5 and 3.

34 Electrons and holes.

35 Releases an electron-hole pair creating voltage.

36 Reduce the depletion layer.

37 Recombine, releasing a quanta of radiation.

38 Semiconductor compounds.

PARTICLES AND WAVES

1 2 (1×10^2 m)

2 Same mass

3 A positive electron

4 Quarks and leptons

5 Quarks

6 Baryons 3, mesons 2

7 $+\frac{2}{3}, -\frac{1}{3}$

8 2 up and 1 down. $(2 \times +\frac{2}{3}) + (1 \times -\frac{1}{3}) = +1$

9 Bosons (gluons, W- and Z-bosons, and photons)

10 The Higgs boson (after Prof. Peter Higgs)

11 In an electric field, an electric charge experiences a force.

12 An electric field applied to a conductor causes the free electric charges in it to move.

13 Work W is done when a charge Q is moved in an electric field.

14 The potential difference between two points is a measure of the work done in moving one coulomb of charge between the two points.

15 If one joule of work is done moving 1 coulomb of charge between two points, the potential difference between the two points is 1 volt.

16 9 V

17 1.88×10^7 m s^{-1}

18 A magnetic field exists round a wire carrying moving charges.

19 The charge will experience a force perpendicular to both the velocity and the magnetic field.

20 A synchrotron.

21 Rutherford showed that a the nucleus has a relatively small diameter compared with that of the atom b most of the mass of the atom is concentrated in the nucleus. See also pp. 92–93.

22 See p 94

23 In fusion two nuclei combine to form a nucleus of larger mass number.

24 See p 94

25 In fission a nucleus of large mass number splits into two nuclei of smaller mass numbers, usually with the release of neutrons. Fission may be spontaneous or induced by neutron bombardment.

26 The frequency of a wave is the same as the frequency of the source producing it.

27 Interference.

28 Reflection, refraction, diffraction and interference are characteristic behaviours of all types of waves.

29 500 nm.

30 red 644 nm green 509 nm blue 480 nm

31 The energy of a wave depends on its amplitude.

32 The amplitudes are in step.

33 Coherent waves have the same frequency, with a constant phase relationship (either they stay in phase or have a constant phase difference).

34 See p 102

35 Photoelectric emission from a surface occurs only if the frequency of the incident radiation is greater than some threshold frequency f_0 which depends on the nature of the surface. For frequencies smaller than the threshold value, an increase in the irradiance of the radiation at the surface will not cause photoelectric emission.

36 For frequencies greater than the threshold value, the photoelectric current produced by monochromatic radiation is directly proportional to the irradiance of the radiation at the surface.

37 A beam of radiation can be regarded as a stream of individual energy bundles called photons, each having an energy dependent on the frequency of the radiation.

38 4×10^{-19} J.

39 Photoelectrons are ejected with a maximum kinetic energy which is given by the difference between the energy of the incident photon and the work function of the surface.

40 The ratio $\frac{\sin\theta_1}{\sin\theta_2}$ is a constant when light passes obliquely from medium 1 to medium 2. The absolute refractive index, n, of a medium is the ratio $\frac{\sin\theta_1}{\sin\theta_2}$ where θ_1 is in a vacuum (or air as an approximation) and θ_2 is in the medium.

41 1·79

42 1.68×10^8 m s^{-1}

43 See pp. 110–111

44 The refractive index depends on the frequency of the incident light.

45 See pp. 110–111

46 The irradiance at a surface on which radiation is incident is the power per unit area.

47 Electrons in a free atom occupy discrete energy levels.

48 See p 113

49 Spontaneous emission of radiation is a random process (analogous to the radioactive decay of nucleus) when an electron drops from an excited state and emits a photon.

50 See p 115

INDEX